RESIDENTIAL STEEL FRAMING
CONSTRUCTION GUIDE

©1994 TECHNICAL PUBLICATIONS

Dedicated to:
Troy, Travis, Trevor and Kassandra

TECHNICAL PUBLICATIONS
3020 Builders Ave.
Las Vegas, Neveda 89101

(702) 598-4365   FAX (702) 598-1733

Manufactured in the United States of America

TABLE OF CONTENTS

A. FLOOR SYSTEM

	A.1	TYPICAL FLOOR FRAMING SCHEMATICS
	A.2	FLOOR JOIST TO TRACK - BEARING ON FOUNDATION
	A.3	PARALLEL FLOOR JOISTS AT FOUNDATION
	A.4	FLOOR JOIST AT FOUNDATION WALL POCKET
	A.5 -7	FLOOR JOIST FLUSH WITH TOP OF FOUNDATION
	A.8	FLOOR JOIST SPLICE OVER STEEL OR BUILT-UP BEAM
	A.9	CONTINUOUS FLOOR JOIST OVER STEEL OR BUILT-UP BEAM
	A.10 - 11	FLOOR JOIST FRAMED FLUSH TO STEEL OR BUILT-UP BEAM
	A.12	FLOOR JOISTS AT SUNKEN FLOOR
	A.13	FLOOR JOIST SPLICE OVER INTERIOR LOAD BEARING STUD WALL
	A.14	CONTINUOUS FLOOR JOIST OVER LOAD BEARING STUD WALL
	A.15	FLOOR JOIST TO EXTERIOR WALL – LOAD BEARING
	A.16	FLOOR JOIST PARALLEL TO EXTERIOR WALL
	A.16a	FLOOR JOIST PARALLEL TO EXTERIOR WALL (ALTERNATE)
	A.17	CANTILEVERED FLOOR JOIST AT BRICK VENEER
	A.18	CANTILEVERED FLOOR JOIST AT FLUSH BALCONY FLOOR
	A.19	CANTILEVERED FLOOR AT STEP DOWN BALCONY FLOOR
	A.20	CANTILEVERED FLOOR AT WOOD DECK BALCONY
	A.21	OPENING IN FLOOR JOISTS
	A.22	JOIST HEADER TO BUILT-UP JOISTS
	A.23	JOIST HEADER TO FLOOR JOIST
	A.24	HEEL CUT FLOOR JOIST
	A.25	JOISTS SUPPORTED AT BEARING STUDS
	A.26	FLOOR JOIST CONNECTION TO INTERIOR STEM WALL

B. WALL FRAMING

	B.1a	TYPICAL WALL FRAMING ELEVATION - 2 STORY
	B.1b	TYPICAL WALL FRAMING ELEVATION
	B.2	WINDOW OPENING GREATER THAN 4 FEET WIDE – NON-LOAD BEARING
	B.3	WINDOW OPENING LESS THAN 4 FEET WIDE – NON-LOAD BEARING
	B.4	WINDOW OPENING GREATER THAN 4 FEET WIDE – LOAD BEARING
	B.5	WINDOW OPENING LESS THAN 4 FEET WIDE – LOAD BEARING
	B.6	DOOR OPENING LESS THAN 4 FEET WIDE – NON-LOAD BEARING
	B.7	DOOR OPENING GREATER THAN 4 FEET WIDE – NON-LOAD BEARING
	B.8	DOOR OPENING LESS THAN 4 FEET WIDE – LOAD BEARING
	B.9	DOOR OPENING GREATER THAN 4 FEET WIDE – LOAD BEARING
	B.10	JAMB AT TOP OF WALL
	B.11	JAMB AT BOTTOM OF WALL
	B.12	HEADER TO JAMB STUD DETAILS
	B.13	OPENING SILL DETAIL – SINGLE TRACK
	B.14	OPENING HEAD DETAIL – SINGLE TRACK WITH HEADER
	B.15	TOP OF NON-LOAD BEARING WALL TO PARALLEL FLOOR JOISTS
	B.16	OPENING SILL DETAIL – BUILT-UP MEMBERS
	B.17	OPENING HEAD DETAIL – LOAD BEARING JAMB AND HEAD
	B.18	JAMB AT FLOOR JOISTS
	B.19	TYPICAL STUD TO SILL TRACK CONNECTION
	B.20	PARTY WALL AT LOAD BEARING WALLS
	B.21	NON-LOAD BEARING SOUND PARTITION DETAILS
	B.22	DOOR JAMB BASE AT FRAMING
	B.23	DOOR JAMB BASE AT SLAB ON GRADE

C. CEE CHANNEL ROOF SYSTEM

C.1	TYPICAL ROOF FRAMING PLAN
C.2a	TYPICAL RAFTER FRAMED ROOF SECTION
C.2b	TYPICAL RAFTER FRAMED VAULTED / CATHEDRAL CEILING
C.3	TRUSS EAVE DETAIL
C.4 - 8	RAFTER EAVE DETAIL
C.9	RIDGE BOARD DETAIL
C.10	RIDGE DETAIL
C.11	COLLAR TIE AT RAFTER DETAIL
C.12	RAFTER TO DIAGONAL BRACE DETAIL
C.13a	TYPICAL KING POST TRUSS PROFILE
C.13b	TYPICAL SCISSORS TRUSS PROFILE
C.14 - 15	WEB TO BOTTOM CHORD DETAIL
C.16	WEB TO TOP CHORD DETAIL
C.17	WEB AT PEAK OF TRUSS DETAIL
C.18	TRUSS END AT EXTERIOR WALL
C.19	SCISSORS TRUSS END AT EXTERIOR WALL
C.20 - 22	GABLE ROOF END DETAIL
C.21 - 23	BALLOON FRAMED GABLE ROOF END DETAIL
C.24	TRACK TO TRACK DETAIL
C.25	DORMER RAFTER AT ROOF RAFTER DETAIL
C.26	DORMER RIDGE AT MAIN ROOF DETAIL
C.27	VALLEY FLASHING DETAIL
C.28	RAFTER WITH REDUCED SIZE OVERHANG
C.29	WOODTAIL CONNECTION TO TRUSS

D. GUS TRUSS™ ROOF SYSTEM

D.1a - 1b	TYPICAL GUS TRUSS™ ELEVATION
D.2	GUS TRUSS™ MEMBER SECTIONS
D.3	TRUSS WEB CONNECTION DETAIL
D.4a	KING POST DETAIL
D.4b	KING POST WITH GUSSET DETAIL
D.5	SCISSORS TRUSS WITH CLIPPED CEILING DETAIL
D.6	STEP DOWN TOP CHORD DETAIL
D.7	SCISSORS TRUSS BOTTOM CHORD DETAIL
D.8	COMPOSITE TRUSS DETAIL (SCISSORS / COMMON)
D.9	OVERHANG DETAIL – FLAT BOTTOM CHORD
D.10	OVERHANG DETAIL – SCISSORS TRUSS
D.11	ZERO OVERHANG DETAIL – FLAT FASCIA
D.12	ZERO OVERHANG DETAIL – RAKED FASCIA
D.13a -13b	TRUSS CONNECTION TO FACE OF STUD
D.14	TRUSS CONNECTION TO HEADER – ZERO OVERHANG
D.15	TRUSS CONNECTION TO HEADER – RAKED FASCIA
D.16	JACK TRUSS CONNECTION TO GIRDER TRUSS
D.17	COMMON TRUSS CONNECTION TO GIRDER TRUSS
D.18	BLOCKED TRUSS HEEL SECTION
D.19	TYPICAL TRUSS GUSSET
D.20	TRUSS BLOCKING DETAIL
D.21	GUS TRUSS™ TO TOP TRACK DETAIL
D.21a - 21b	BOTTOM CHORD TO TOP PLATE CONNECTION DETAIL
D.22	TYPICAL HIP ROOF PLAN
D.23	TYPICAL OVERFRAME (CALIFORNIA FRAMING) TRUSSES
D.23a	TYPICAL OVERFRAME DETAIL
D.24	TYPICAL SOFFIT FRAMING DETAIL
D.25	SECTION AT RAKE WALL TRUSS
D.26	DRAG TRUSS TO SHEAR WALL TOP TRACK
D.27	ROOF TRUSS CONNECTION AT PARTY WALL
D.28	WOOD RAFTER TAIL CONNECTION TO GUS TRUSS™
D.29	SECTION AT RAKE WALL TRUSS WITH WOOD OUTRIGGERS
D.30	SECTION AT BALLOON FRAMED RAKE WALL WITH WOOD OUTRIGGERS
D.31	SECTION AT BALLOON FRAMED RAKE WALL OVER OPENING
D.32a	TOP CHORD BEARING FLOOR TRUSS DETAIL
D.32b	TOP CHORD BEARING FLOOR TRUSS DETAIL (ALTERNATE)

E. DETAILS AT EXTERIOR WALLS

	E.1	BASE OF WALL AT SLAB ON GRADE
	E.2	FLOOR JOISTS BEARING ON FOUNDATION
	E.3	FLOOR JOISTS PARALLEL TO FOUNDATION
	E.4	FLOOR JOIST SUPPORT AT CONTINUOUS WALL
	E.5	FLOOR JOISTS PARALLEL TO CONTINUOUS WALL
	E.6	FLOOR FRAMING AT EXTERIOR WALL
	E.7	FLOOR JOISTS PARALLEL TO EXTERIOR WALL
	E.7a	FLOOR JOISTS PARALLEL TO EXTERIOR WALL (ALTERNATE)
	E.8	FLOOR CANTILEVER
	E.9	BALCONY WITH STEP DOWN
	E.10	WOOD DECK BALCONY
	E.11	ROOF EAVE
	E.12	ROOF TRUSS EAVE
	E.13	ROOF EAVE AND SOFFIT
	E.14	ROOF EAVE AT CATHEDRAL CEILING
	E.15	ROOF TRUSS BEARING
	E.16	ROOF SCISSORS TRUSS BEARING
	E.17	ROOF GABLE END
	E.18	ROOF GABLE END AT CATHEDRAL
	E.19	CANTILEVERED ROOF GABLE END
	E.20	CANTILEVERED GABLE END AT CATHEDRAL
	E.21a	TYPICAL SHEAR PANEL ELEVATION
	E.21b	TYPICAL SHEAR PANEL ELEVATION – 2 STORY
	E.22	STACKED SECOND FLOOR SHEAR WALL WITH COMMON GUSSET PLATE
	E.23	SECOND FLOOR SHEAR WALL STRAP TIE HOLDOWN DETAIL
	E.23a	SECOND FLOOR SHEAR WALL STRAP TIE HOLDOWN DETAIL (ALTERNATE)
	E.24	SECOND FLOOR SHEAR WALL HOLDOWN DETAIL
	E.25	SHEAR WALL GUSSET PLATE AND HOLDOWN ASSEMBLY
	E.26	TYPICAL TRACK ANCHORAGE DETAIL AT EXTERIOR WALL ON SLAB
	E.27	TYPICAL HOLDOWN DETAIL AND SCHEDULE
	E.28	SECOND FLOOR SHEAR HOLDOWN TO FLOOR JOISTS
	E.29	EXTERIOR WALL SECTION WITH PARALLEL FLOOR JOISTS
	E.30	FLOOR JOIST TO FLUSH FRAMED BEAM CONNECTION
	E.31	TYPICAL DROPPED BEAM TO WALL CONNECTION
	E.32	STEM WALL DETAIL
	E.33	JOIST CONNECTION AT STEM WALL
	E.34	HOLDOWN DETAIL AT STEM WALL
	E.35	PARALLEL FLOOR JOIST TO STEM WALL

F. MISC. BRIDGING, BLOCKING, REINFORCEMENT, ETC.

	F.1	TOP TRACK LOAD DISTRIBUTION DETAILS
	F.2	SHEAR WALL HOLDOWN AT BASE
	F.3	SHEAR WALL HOLDOWN AT SECOND FLOOR
	F.4	SOLID BLOCKING
	F.5	CROSS BRIDGING
	F.6	JOIST AND RAFTER BRIDGING
	F.7	WALL BRIDGING
	F.8	WALL BRIDGING (ALTERNATE)
	F.9a	TOP TRACK SPLICE DETAIL
	F.9b	BOTTOM TRACK SPLICE DETAIL
	F.10	TYPICAL EXTERIOR CORNER FRAMING
	F.10a	EXTERIOR CORNER FRAMING WITH HOLDOWN
	F.10b	TYPICAL INTERIOR CORNER FRAMING
	F.11	TYPICAL INTERIOR INTERSECTION FRAMING
	F.11a	ALTERNATE INTERIOR INTERSECTION FRAMING
	F.12	JOIST, STUD OR RAFTER WEB PENETRATIONS
	F.13	SHEAR PANEL GRACING DETAILS
	F.13a	DIAGONAL STRAP ATTACHMENT TO HEADER
	F.14	WALL HORIZONTAL BLOCKING / BRIDGING DETAIL
	F.15	TYPICAL STAIR STRINGER CONNECTION
	F.16	EXTERIOR POP-OUT DETAIL
	F.17	TYPICAL POT SHELF DETAIL

(continued)

F. MISC. BRIDGING, BLOCKING, REINFORCEMENT, ETC. (continued)

	F.18	TYPICAL PLASTER SOFFIT DETAIL
	F.19	TYPICAL ARCH OPENING DETAIL
	F.20	ATTACHMENT BACKING DETAIL
	F.21	TYPICAL LINTEL BEAM CONNECTION
	F.22	TYPICAL LINTEL BEAM CONNECTION (ALTERNATE)
	F.23	TYPICAL BOXED HEADER AND BEAM DETAILS
	F.24	TYPICAL ELECTRICAL ATTACHMENT DETAIL
	F.25	SHEAR WALL TRACK REINFORCEMENT AND ANCHORAGE
	F.26	DRAG STRUT BLOCKING INTO FLOOR DIAPHRAGM
	F.27	INVERTED CHEVRON TYPE STRAP CONNECTION TO RIM JOIST
	F.28	MECHANICAL UNIT SUPPORT BETWEEN TRUSSES IN ATTIC SPACE

G. FASTENING SCHEDULE RECOMMENDATIONS

	G.1	ROOF DESIGN (COMPARATIVE FRAMING DESIGNS)
	G.2	FLOOR FRAMING
	G.3	WALL FRAMING (LOAD BEARING)
	G.4	WALL FRAMING (NON-LOAD BEARING / DRYWALL)
	G.5	ROOF FRAMING
	G.6	GYPSUM WALL BOARD (SINGLE PLY)
	G.7	PLYWOOD OR O.S.B. SHEATHING (SINGLE PLY)
	G.8	PLYWOOD OR O.S.B. FOR COMBINATION SUBFLOOR

H. FASTENERS

	H.1	PANCAKE (LOW PROFILE) HEAD SCREW
	H.2	HEX WASHER HEAD SCREW
	H.3	PAN HEAD SCREW
	H.4	PAN FRAMING SCREW
	H.5	HEX HEAD SELF-DRILLING #5 SCREW
	H.6	BUGLE HEAD SCREW
	H.7	TRIM HEAD SCREW
	H.8	WAFER HEAD OR WAFER WINGED SCREW
	H.9	FLAT WINGED PHILLIPS SCREW
	H.10	PILOT POINT SCREW
	H.11	SCREW-SHANK NAIL
	H.12	FIBER CEMENT BOARD SCREWS / CEMENT BOARD SCREWS
	H.13	BASIC DRILL CAPACITY RECOMMENDATIONS
	H.14	ALLOWABLE SCREW LOADS
	H.15	MISC. FASTENER INFORMATION

I. HARDWARE / CONNECTORS

	I.1	MUDSILL ANCHOR
	I.2	MONKEY PAW™ ANCHOR BOLT HOLDER
	I.3	ANCHOR BOLT
	I.4	EPOXY TIE
	I.5 - 6	HOLDOWNS
	I.7	TENSION TIES
	I.8 - 9	HANGERS
	I.10	REINFORCING AND SKEWABLE ANGLES
	I.11	ANGLES
	I.12	SEISMIC AND HURRICANE TIES
	I.13	TWIST STRAPS
	I.14	STRAP TIES
	I.15	COILED STRAPS
	I.16	PLYWOOD SHEATHING CLIPS
	I.17	BRIDGING
	I.18	MISC. CHARTS

FOREWORD

It is important for the user to understand that the details contained in this book are intended solely as a general guide with respect to residential steel framing. The details are not intended to be used as final approved engineering details. They should be reviewed and approved by licensed professional engineers and architects before actual implementation.

Details have been presented as generic and do not specifically refer to or recommend one manufacturer over another. One exception is the Gus Truss™, patented by Hemming Technologies.

The various details contained in this book represent standard framing techniques that have been developed and implemented over many years in the residential construction of custom homes, tract (production) homes and multi family units. The user should realize that other techniques are also used in the field of residential steel framing.

Steel framing provides an excellent alternative and opportunity to traditional wood framing and continues to gain momentum and popularity because of its many advantages. Steel framing is the wave of the future.

Finally, the author welcomes comments and suggestions from users as to how this book might be improved, and would appreciate notification of any inadvertent errors.

E.N. Lorre

A. FLOOR SYSTEM

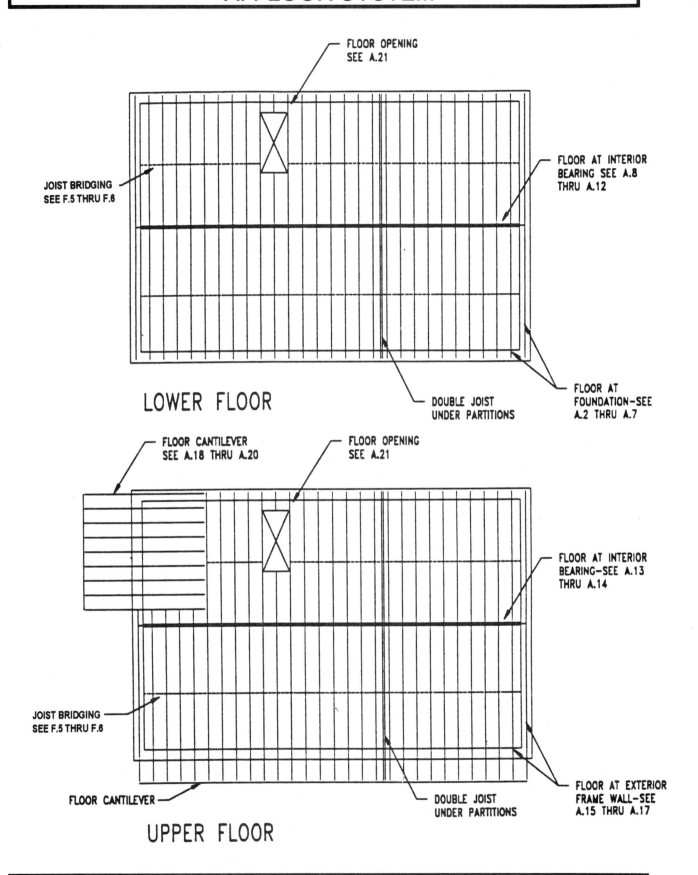

TYPICAL FLOOR FRAMING SCHEMATICS

A. FLOOR SYSTEM

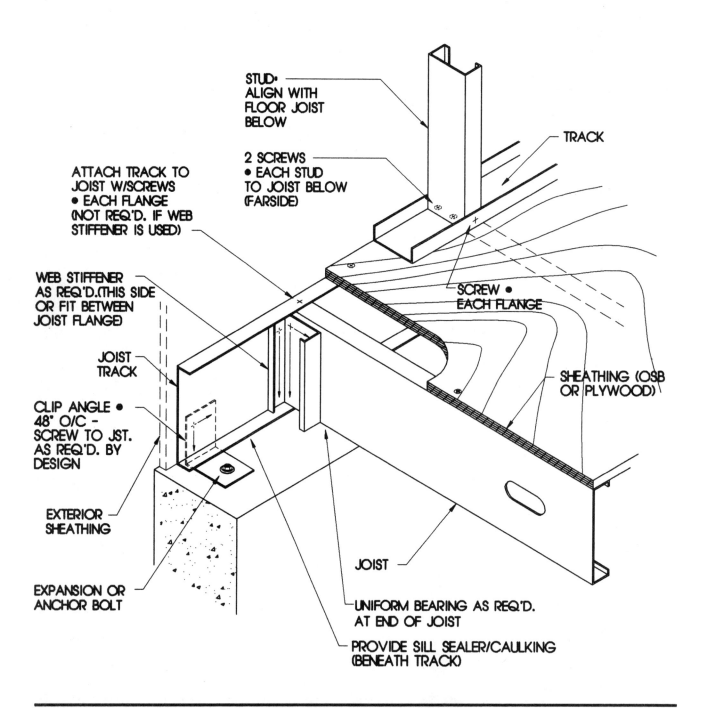

A.2   FLOOR JOIST TO TRACK BEARING ON FOUNDATION

A. FLOOR SYSTEM

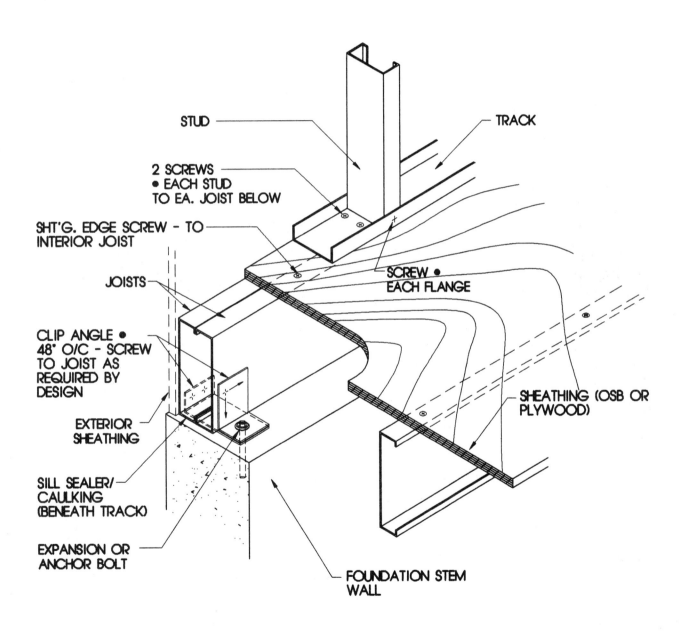

PARALLEL FLOOR JOISTS AT FOUNDATION

A.3

A. FLOOR SYSTEM

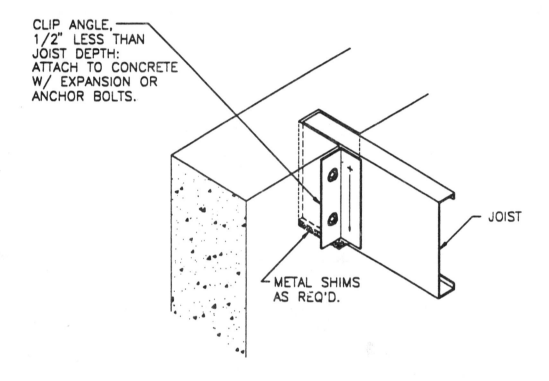

CLIP ANGLE, 1/2" LESS THAN JOIST DEPTH: ATTACH TO CONCRETE W/ EXPANSION OR ANCHOR BOLTS.

JOIST

METAL SHIMS AS REQ'D.

A.4 FLOOR JOIST AT FOUNDATION WALL POCKET

A. FLOOR SYSTEM

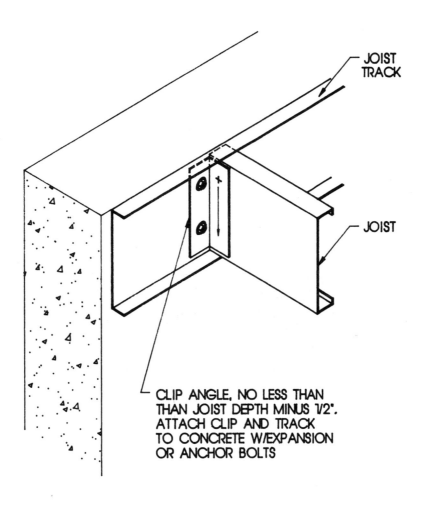

FLOOR JOIST FLUSH WITH TOP OF FOUNDATION

A.5

A. FLOOR SYSTEM

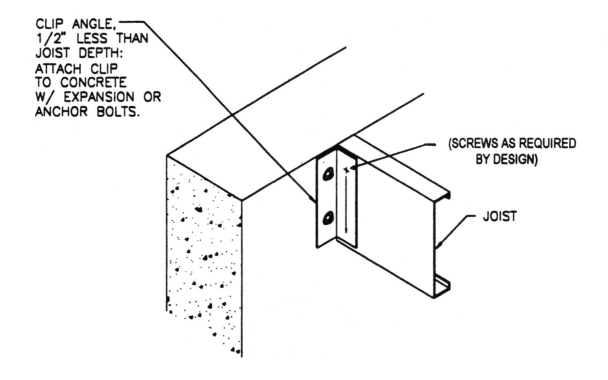

A.6   FLOOR JOIST FLUSH WITH TOP OF FOUNDATION

A. FLOOR SYSTEM

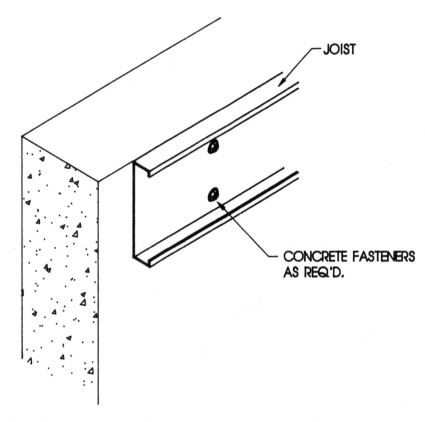

FLOOR JOIST FLUSH WITH TOP OF FOUNDATION

A.7

A. FLOOR SYSTEM

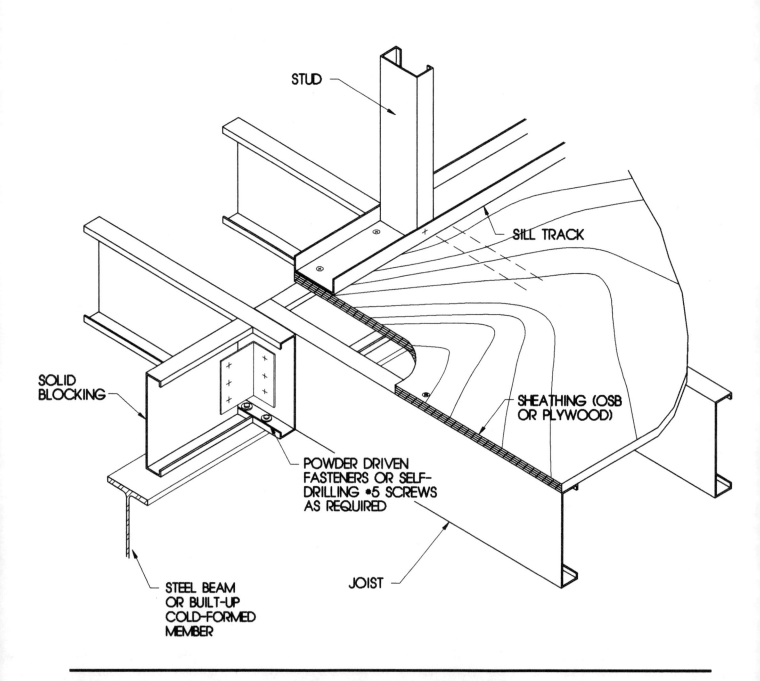

A.8   FLOOR JOIST SPLICE OVER STEEL OR BUILT-UP BEAM

A. FLOOR SYSTEM

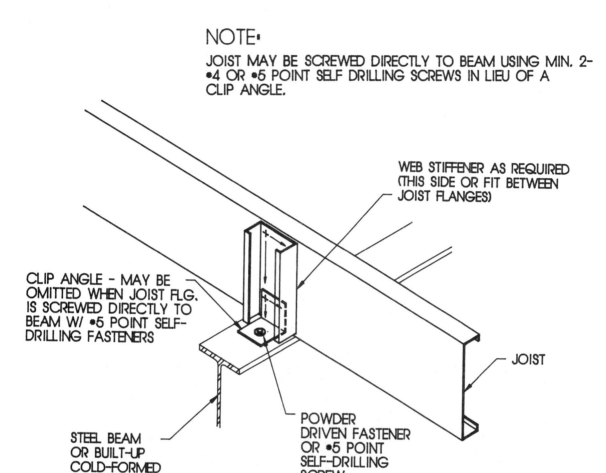

NOTE:
JOIST MAY BE SCREWED DIRECTLY TO BEAM USING MIN. 2- #4 OR #5 POINT SELF DRILLING SCREWS IN LIEU OF A CLIP ANGLE.

WEB STIFFENER AS REQUIRED (THIS SIDE OR FIT BETWEEN JOIST FLANGES)

CLIP ANGLE - MAY BE OMITTED WHEN JOIST FLG. IS SCREWED DIRECTLY TO BEAM W/ #5 POINT SELF-DRILLING FASTENERS

JOIST

STEEL BEAM OR BUILT-UP COLD-FORMED MEMBER

POWDER DRIVEN FASTENER OR #5 POINT SELF-DRILLING SCREW

NOTES:
1. CONTINUOUS BRIDGING REQUIRED BETWEEN EACH JOIST ABOVE BEAM - SEE F.6. SOLID BLOCKING IN EVERY OTHER SPACE MAY BE USED IN LIEU OF BRIDGING.
2. WHEN WALL ABOVE, STUDS MUST ALIGN WITH JOISTS.
3. WEB STIFFENERS ARE NOT REQUIRED WHEN CONTINUOUS SOLID BLOCKING IS USED.

CONTINUOUS FLOOR JOIST OVER STEEL OR BUILT-UP BEAM — **A.9**

A. FLOOR SYSTEM

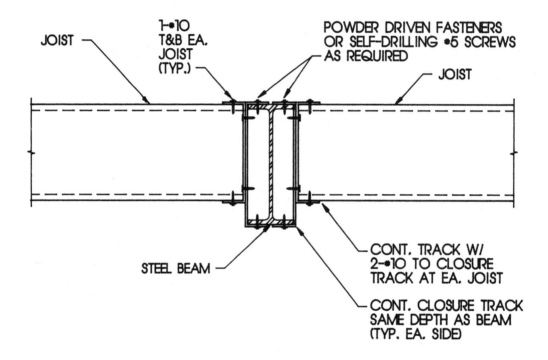

A.10 FLOOR JOIST FRAMED FLUSH TO STEEL OR BUILT-UP BEAM

A. FLOOR SYSTEM

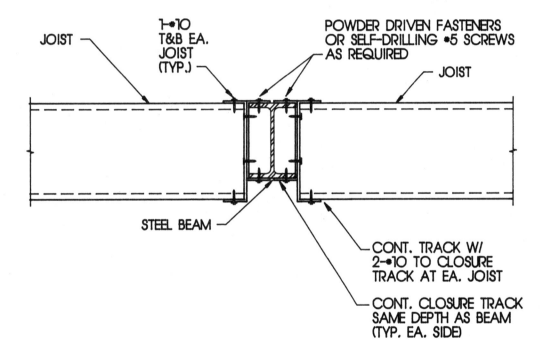

FLOOR JOIST FRAMED FLUSH TO STEEL OR BUILT-UP BEAM

A.11

A. FLOOR SYSTEM

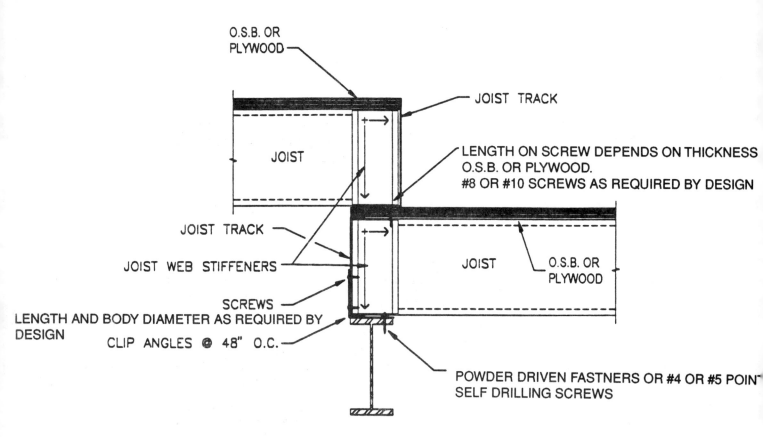

A.12   FLOOR JOISTS AT SUNKEN FLOOR

A. FLOOR SYSTEM

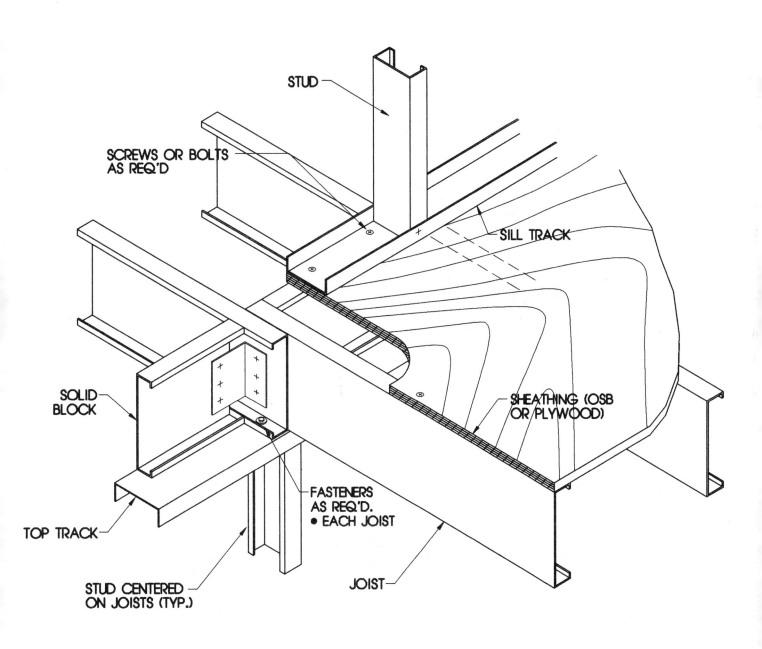

FLOOR JOIST SPLICE OVER INTERIOR LOAD BEARING STUD WALL

A.13

A. FLOOR SYSTEM

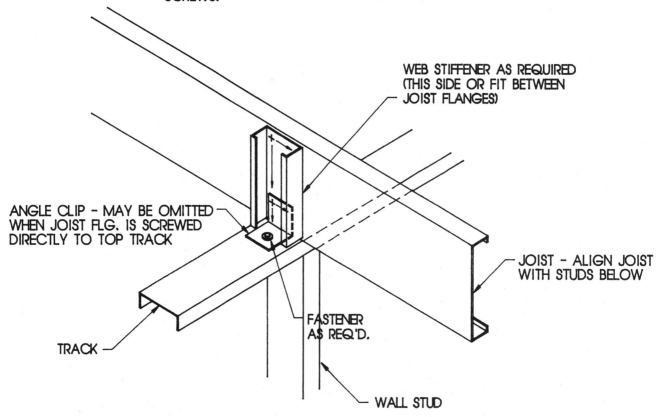

A.14   CONTINUOUS FLOOR JOIST OVER LOAD BEARING STUD WALL

A. FLOOR SYSTEM

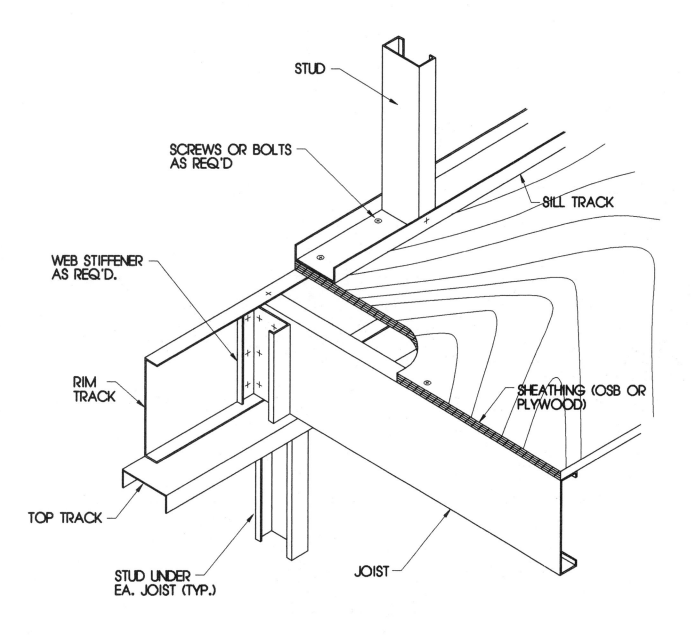

A.15   FLOOR JOIST TO EXTERIOR WALL – LOAD BEARING   **A.15**

A. FLOOR SYSTEM

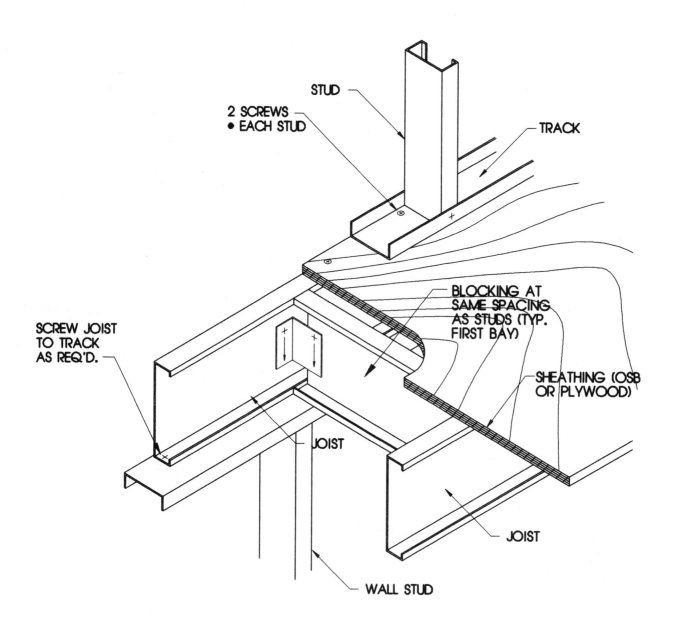

A.16 FLOOR JOIST PARALLEL TO EXTERIOR WALL BEARING ON FOUNDATION

A. FLOOR SYSTEM

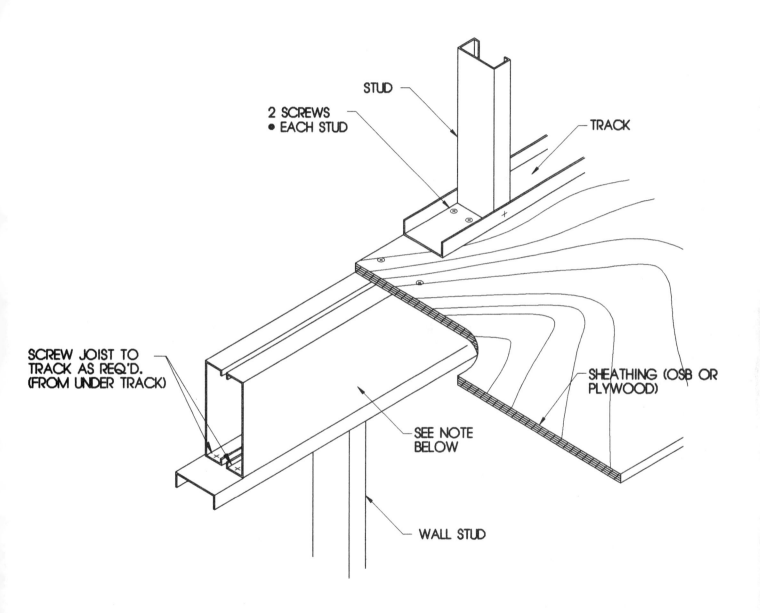

NOTE:
RIM JOIST MAY BE DOUBLED AS SHOWN AND MAY BE UTILIZED TO ELIMINATE THE NEED FOR ADDITIONAL DOOR OR WINDOW HEADERS. THIS DETAIL MAY ALSO BE USED WHERE FIRE RATED WALL CONSTRUCTION IS REQUIRED.

FLOOR JOIST PARALLEL TO EXTERIOR WALL BEARING ON FOUNDATION (ALTERNATE) — **A.16a**

A. FLOOR SYSTEM

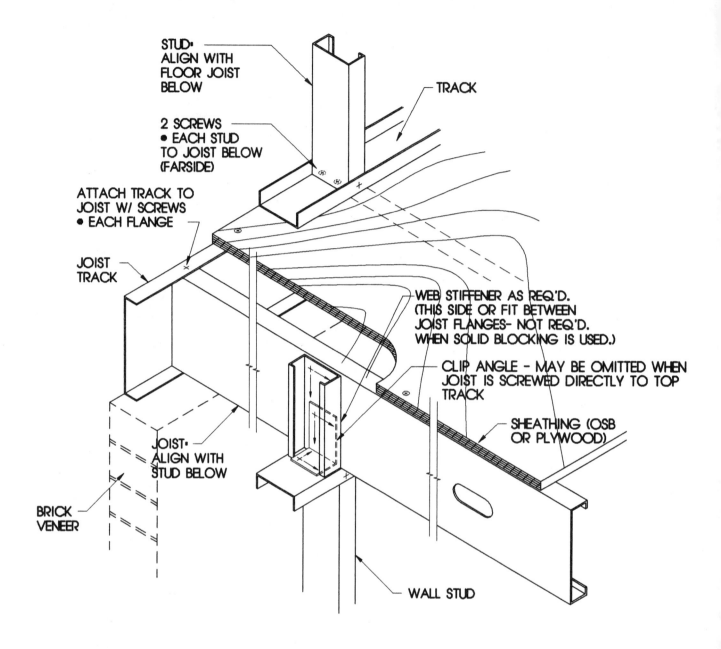

NOTES:
1. PROVIDE CONT. BRIDGING BETWEEN EACH JOIST AT LOWER WALL - SEE F.6.
2. SOLID BLOCKING IN EVERY OTHER SPACE MAY BE USED IN LIEU OF BRIDGING - SEE F.4
3. WHERE AXIAL LOAD BEARING MEMBERS DO NOT ALIGN VERTICALLY PROVIDE DETAIL F.1

A.17   CANTILEVERED FLOOR JOIST AT BRICK VENEER

A. FLOOR SYSTEM

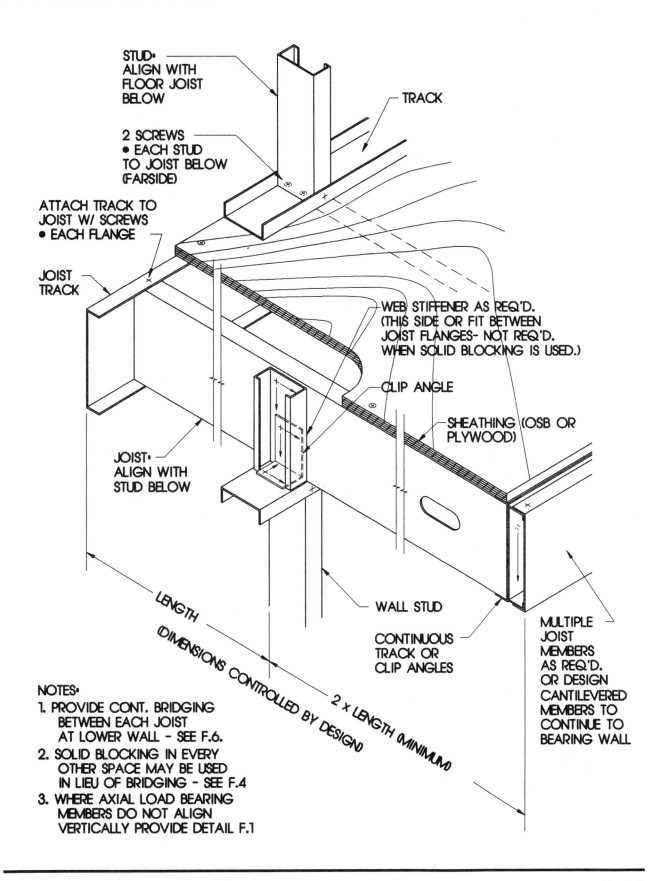

CANTILEVERED FLOOR JOIST AT FLUSH BALCONY FLOOR

A.18

A. FLOOR SYSTEM

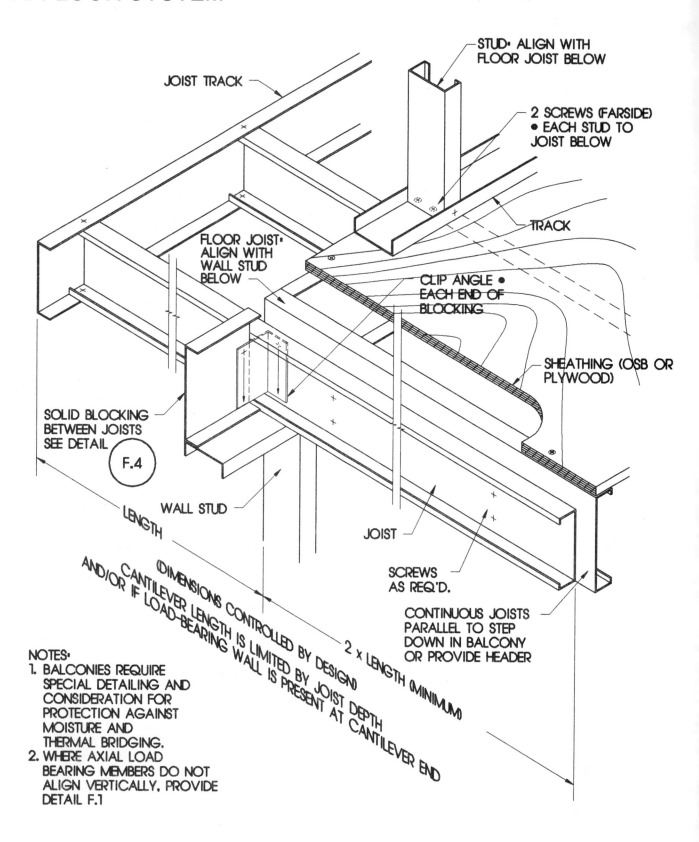

A.19 CANTILEVERED FLOOR AT STEP DOWN BALCONY FLOOR

A. FLOOR SYSTEM

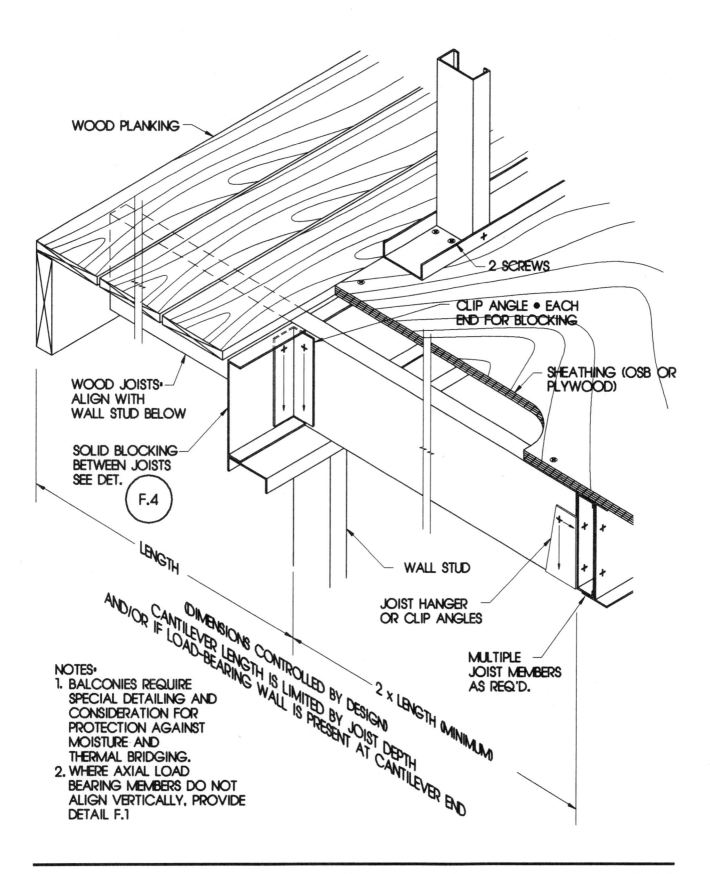

CANTILEVERED FLOOR AT
WOOD DECK BALCONY

A.20

A. FLOOR SYSTEM

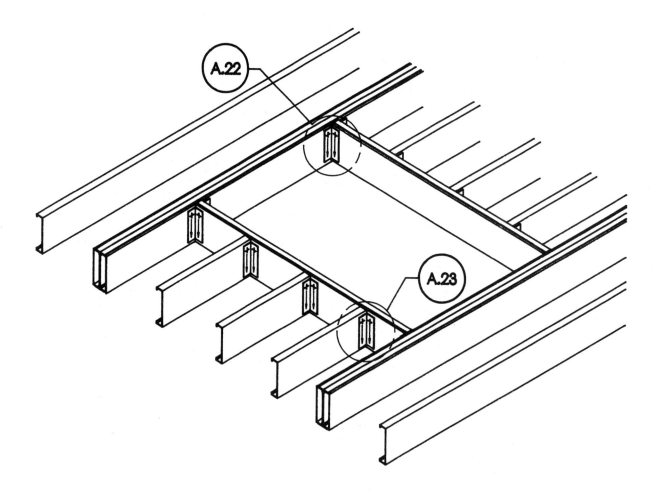

A.21 OPENING IN FLOOR JOISTS

A. FLOOR SYSTEM

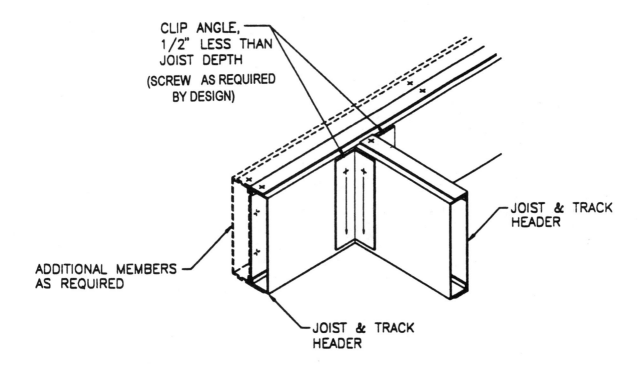

JOIST HEADER TO BUILT-UP JOISTS

A.22

A. FLOOR SYSTEM

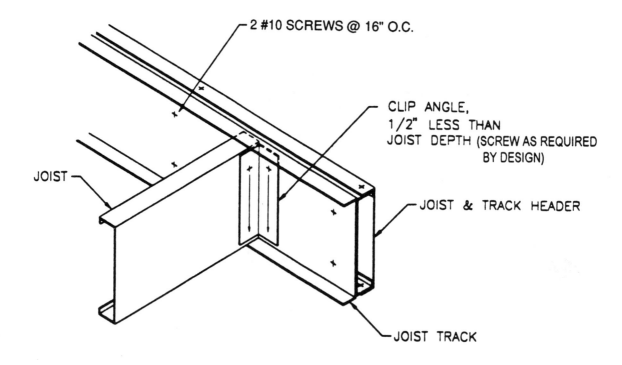

A.23   JOIST HEADER TO FLOOR JOISTS

A. FLOOR SYSTEM

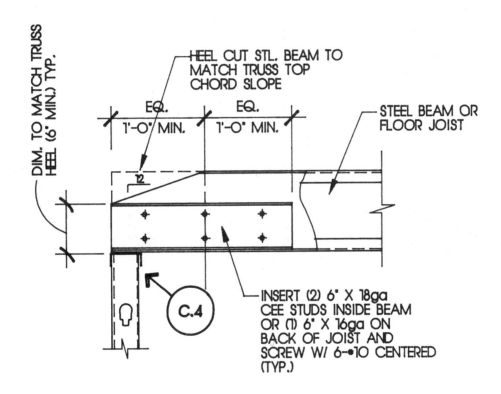

HEEL CUT FLOOR JOIST

A.24

A. FLOOR SYSTEM

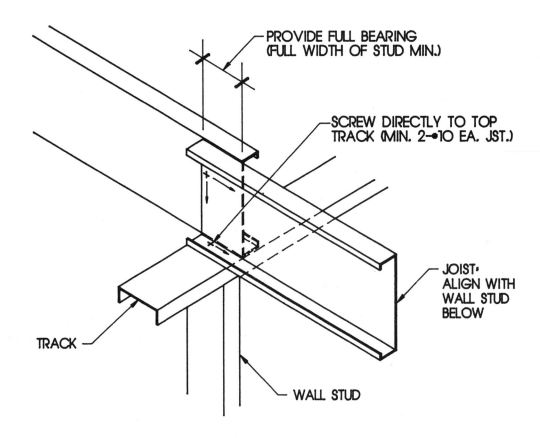

NOTES:
1. CONTINUOUS BRIDGING REQ'D. BETWEEN EACH JOIST ABOVE WALL - SEE F.6. SOLID BLOCKING IN EVERY OTHER SPACE MAY BE USED IN LIEU OF BRIDGING.
2. WHEN WALL ABOVE, STUDS MUST ALIGN WITH JOISTS.
3. WHERE AXIAL LOAD BEARING MEMBERS DO NOT ALIGN VERTICALLY, PROVIDE DETAIL F.1

A.25 JOISTS SUPPORTED AT BEARING STUDS

E. DETAILS AT EXTERIOR WALLS

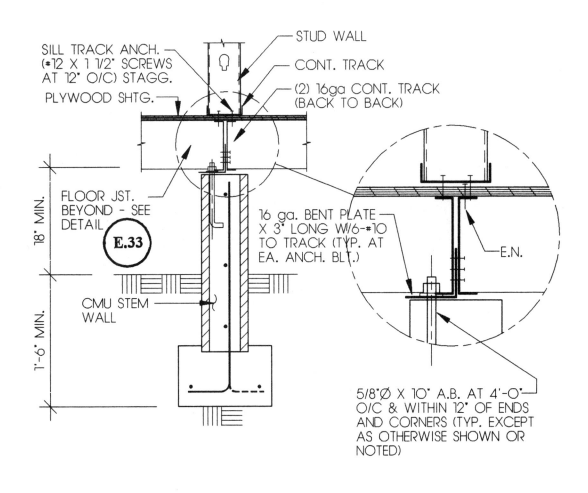

FLOOR JOIST CONNECTION TO INTERIOR STEM WALL

A.26

B. WALL FRAMING

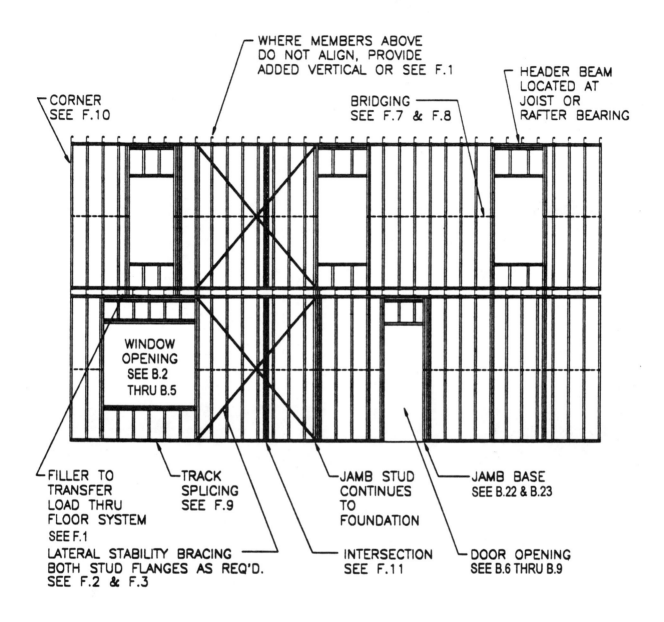

NOTES:
1. JOISTS ALIGN OVER WALL STUDS (TYP.).
2. JAMB MEMBERS MUST BE CARRIED DOWN ALL WALLS TO FOUNDATION (TYP.).
3. STUD WEB PENETRATIONS, SEE F.12.
4. HEADERS FOR OPENINGS MAY BE LOCATED DIRECTLY ABOVE OPENING OR AT JOIST BEARING. WHEN LOCATED AT WINDOW HEAD, CRIPPLE STUDS MUST BE TIGHTLY SEATED FOR FULL BEARING.

TYPICAL WALL FRAMING ELEVATION – 2 STORY **B.1a**

B. WALL FRAMING

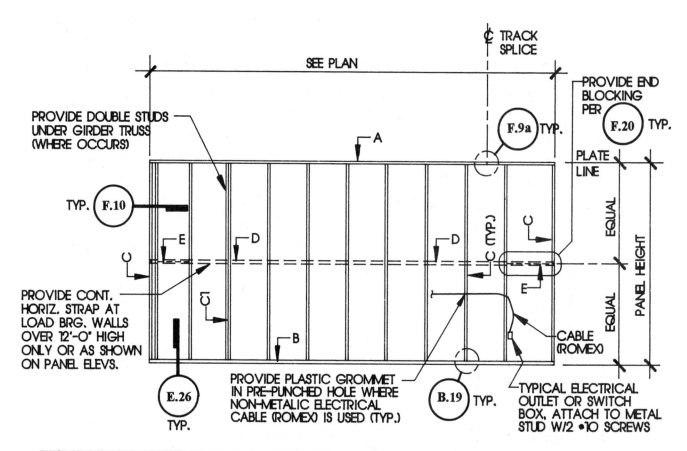

SCHEDULE			
MARK	DESCRIPTION	QTY	SIZE (*)
A	TOP TRACK	-	3 1/2" x 20ga
B	BOTTOM TRACK	-	3 1/2" x 20ga
C	STUD	-	3 1/2" x 20ga
C1	DOUBLE STUD	-	(2)3 1/2" x 20ga
D	STRAP	-	2" x 16 GA.
E	BLOCKING	-	3 1/2" x 20ga

NOTES:

* MEMBER SIZES SHOWN IN THIS DETAIL ARE TYPICAL EXCEPT AS OTHERWISE SHOWN ON THE PLANS OR SPECIFIC PANEL ELEVATIONS.

ALL STUDS SHALL BE SPACED AT 24" O/C EXCEPT AS SHOWN OTHERWISE AND AS NOTED BELOW.

LOAD BEARING STUDS SHALL BE SPACED SO AS TO FALL DIRECTLY UNDER ROOF TRUSSES/RAFTERS OR UNDER FLOOR JOISTS.

B.1b   TYPICAL WALL FRAMING ELEVATION

B. WALL FRAMING

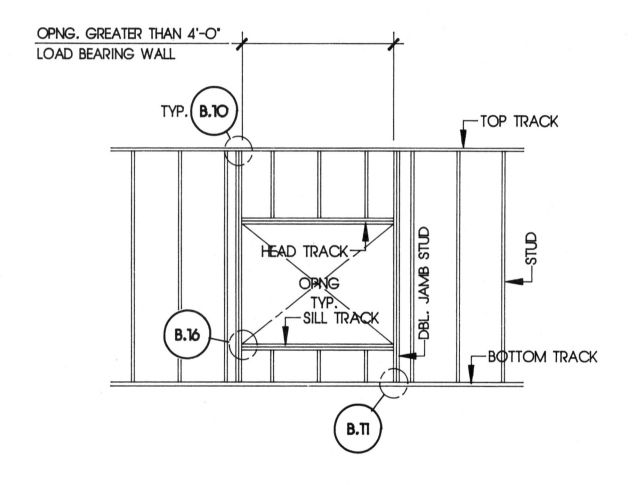

WINDOW OPENING GREATER THAN 4 FEET WIDE – NON-LOAD BEARING

B.2

B. WALL FRAMING

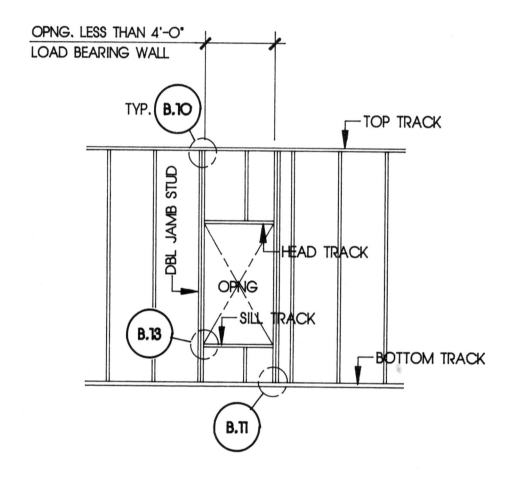

B.3 WINDOW OPENING LESS THAN 4 FEET WIDE – NON-LOAD BEARING

B. WALL FRAMING

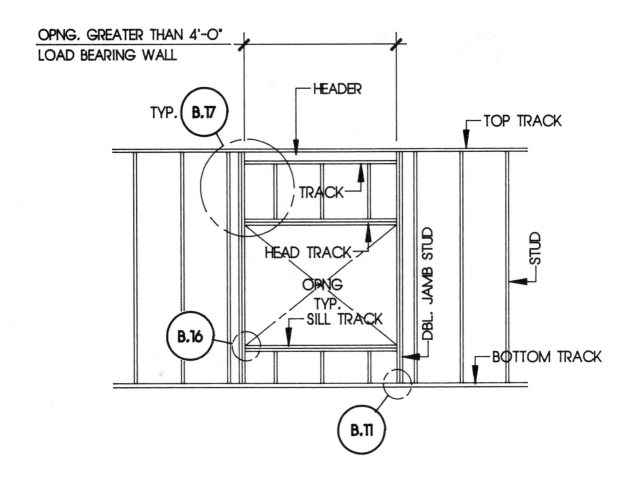

WINDOW OPENING GREATER THAN 4 FEET WIDE – LOAD BEARING

B.4

B. WALL FRAMING

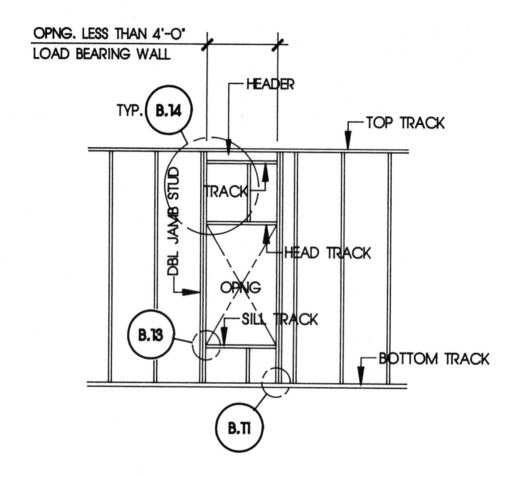

B.5 WINDOW OPENING LESS THAN 4 FEET WIDE LOAD BEARING

B. WALL FRAMING

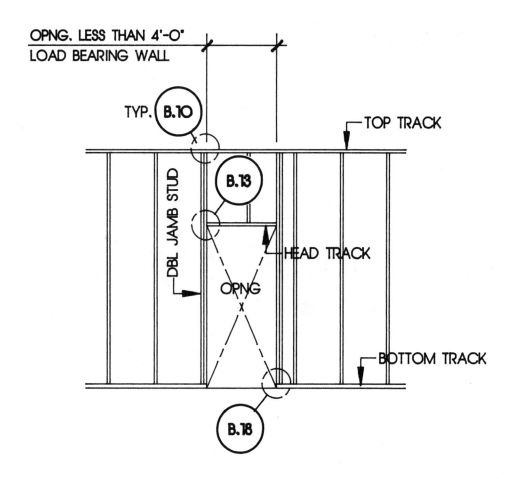

**DOOR OPENING LESS THAN
4 FEET WIDE – NON-LOAD BEARING**

B.6

B. WALL FRAMING

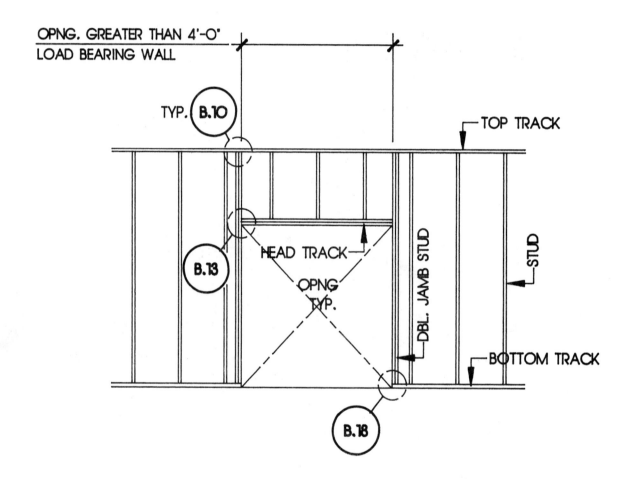

B.7   DOOR OPENING GREATER THAN 4 FEET WIDE – NON-LOAD BEARING

B. WALL FRAMING

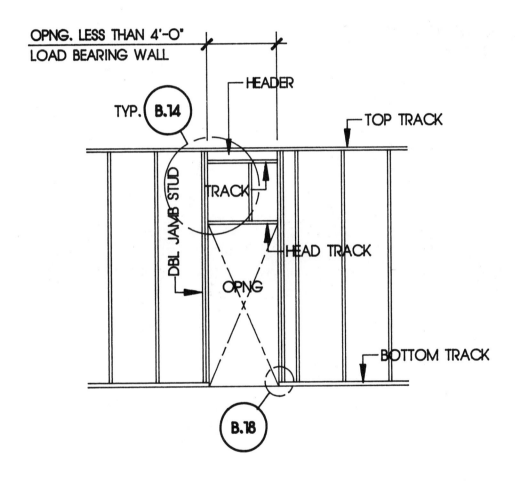

**DOOR OPENING LESS THAN
4 FEET WIDE – LOAD BEARING**

B.8

B. WALL FRAMING

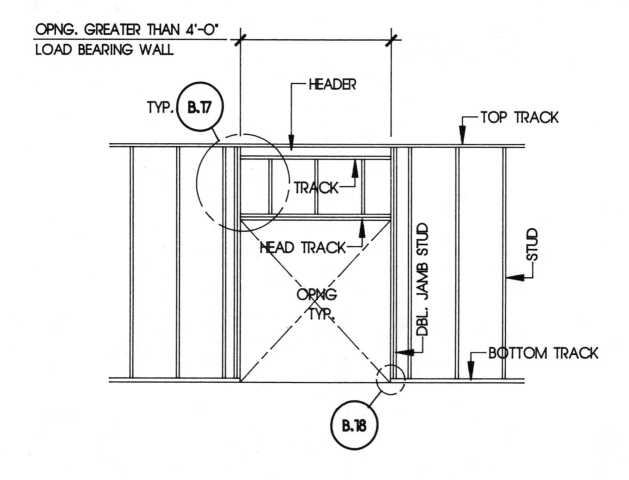

B.9 DOOR OPENING GREATER THAN
4 FEET WIDE – LOAD BEARING

B. WALL FRAMING

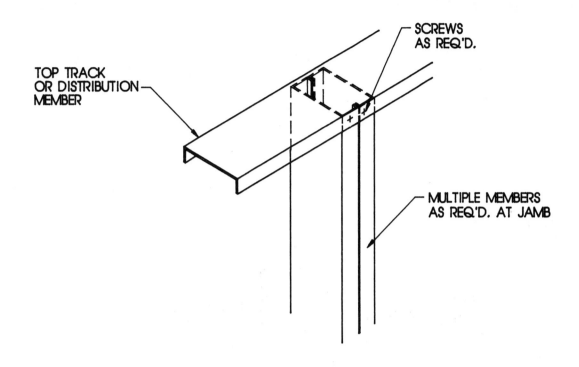

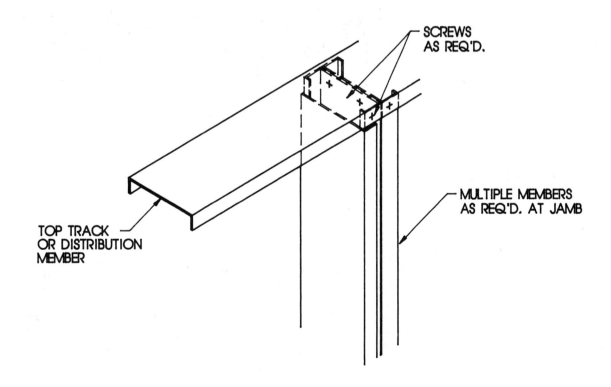

JAMB AT TOP OF WALL

B.10

B. WALL FRAMING

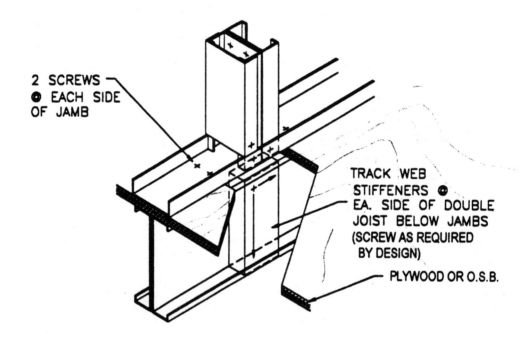

NOTE:
WEB STIFFENERS MAY NOT BE NEEDED DEPENDING UPON THE PARTICULAR DESIGN REQUIREMENTS.

B.11   JAMB AT BOTTOM OF WALL

B. WALL FRAMING

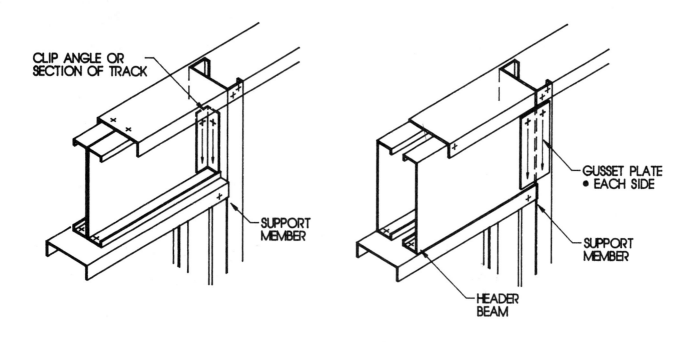

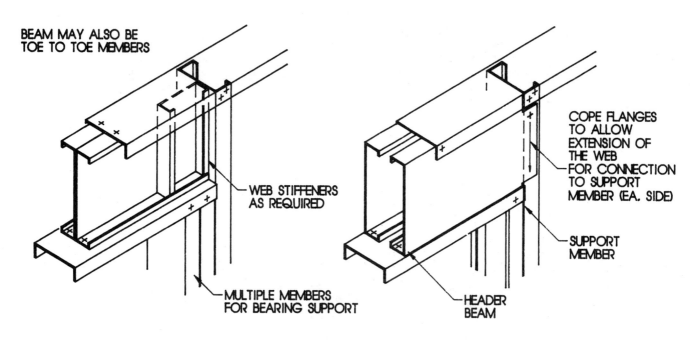

HEADER TO JAMB STUD DETAILS   **B.12**

B. WALL FRAMING

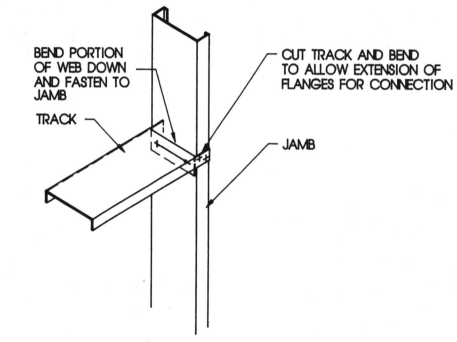

B.13 OPENING SILL DETAIL – SINGLE TRACK

B. WALL FRAMING

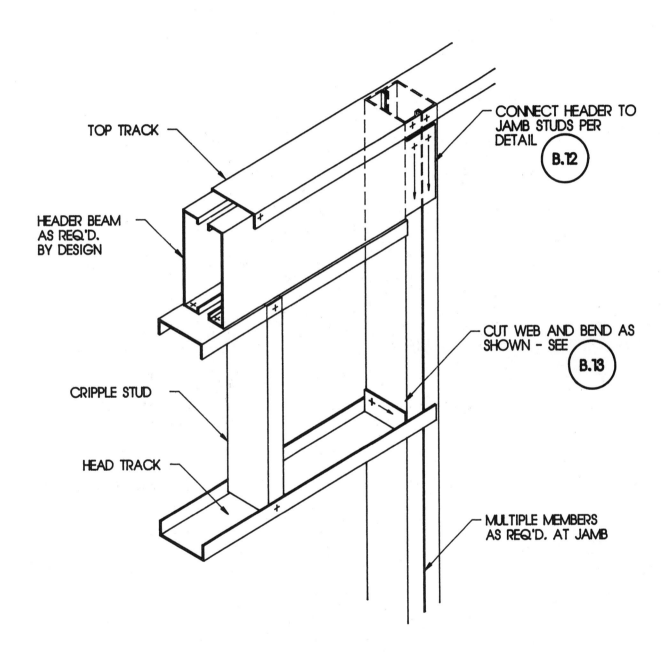

OPENING HEAD DETAIL – SINGLE TRACK WITH HEADER

B.14

B. WALL FRAMING

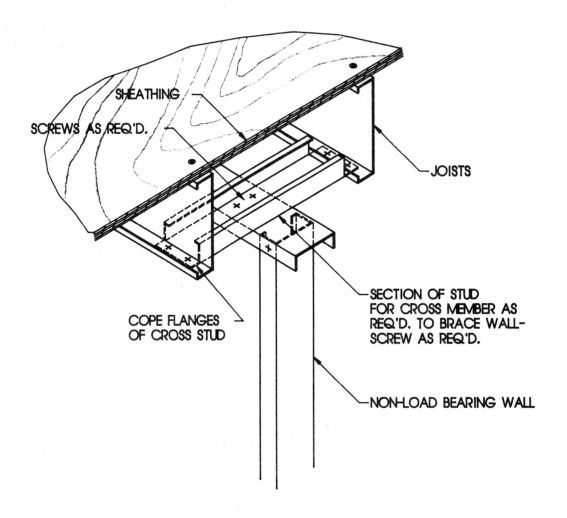

B.15   TOP OF NON-LOAD BEARING WALL TO PARALLEL FLOOR JOISTS

B. WALL FRAMING

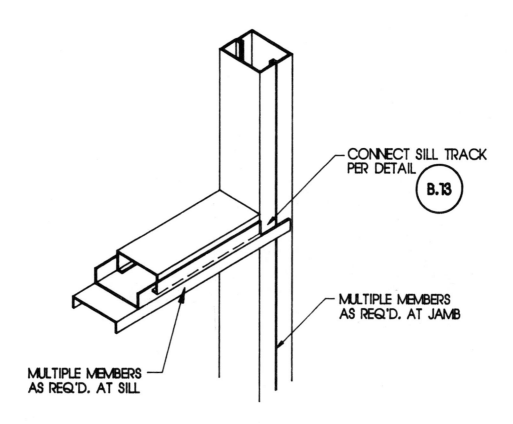

OPENING SILL DETAIL – BUILT-UP MEMBERS B.16

B. WALL FRAMING

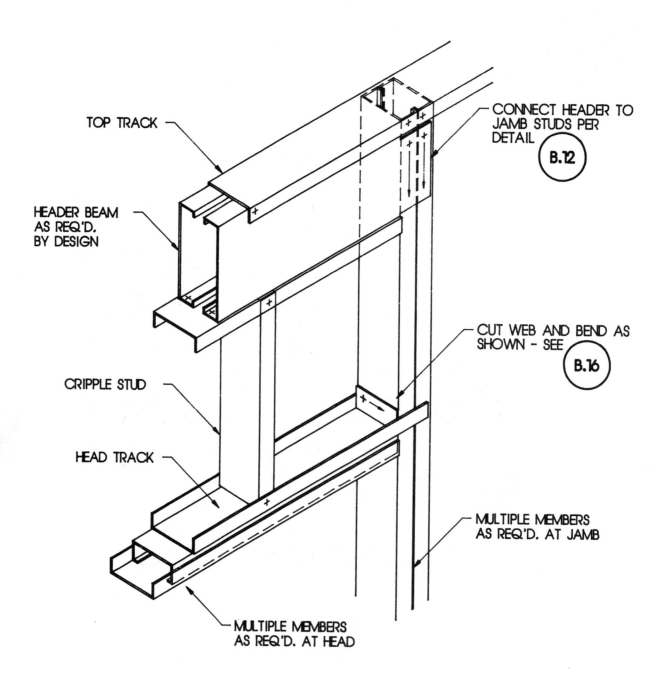

B.17 OPENING HEAD DETAIL – LOAD BEARING JAMB AND HEAD

B. WALL FRAMING

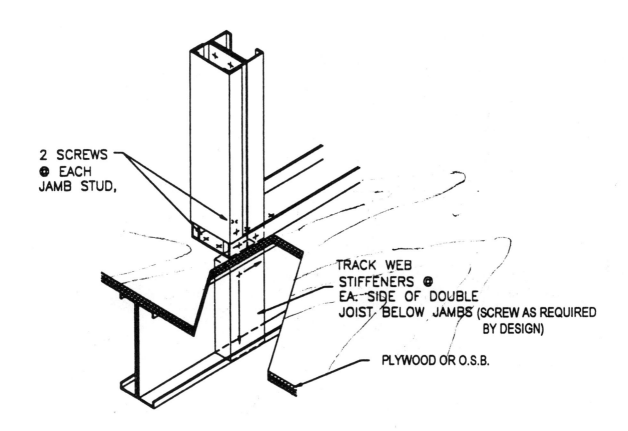

- 2 SCREWS @ EACH JAMB STUD.
- TRACK WEB STIFFENERS @ EA. SIDE OF DOUBLE JOIST BELOW JAMBS (SCREW AS REQUIRED BY DESIGN)
- PLYWOOD OR O.S.B.

NOTE:
WEB STIFFENERS MAY NOT BE NEEDED DEPENDING UPON THE PARTICULAR DESIGN REQUIREMENTS.

JAMB AT FLOOR JOISTS — B.18

B. WALL FRAMING

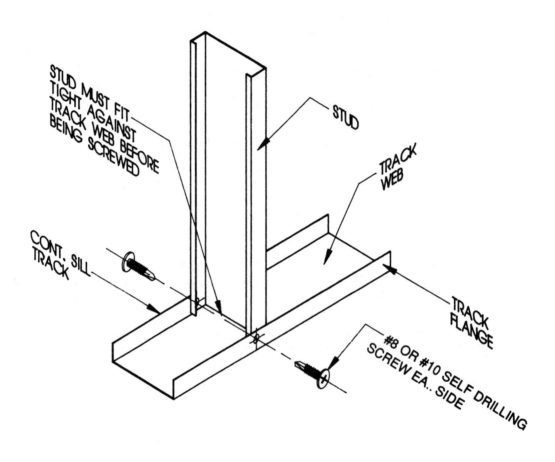

B.19 TYPICAL STUD TO SILL TRACK CONNECTION

B. WALL FRAMING

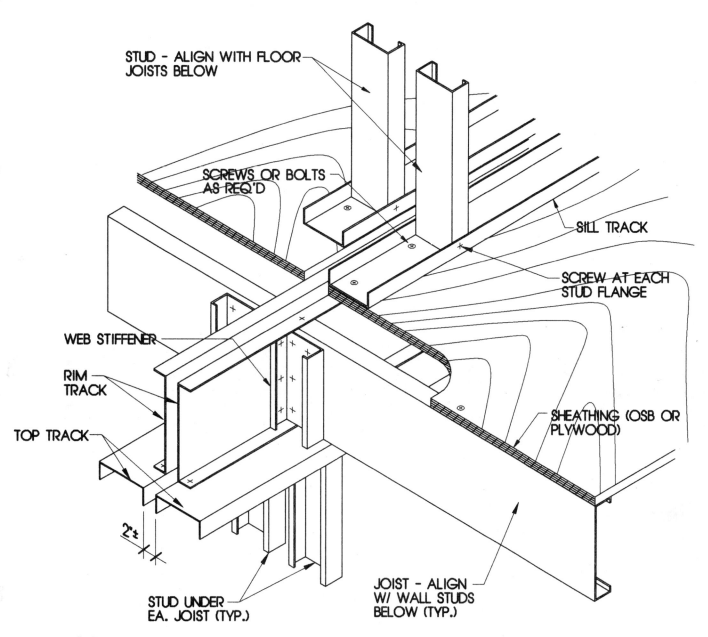

NOTE:
IN ORDER TO FACILITATE THE ATTACHMENT OF DRYWALL WHICH MUST EXTEND TO THE UNDERSIDE OF FLOOR SHEATHING, BLOCKING MUST BE PROVIDED BETWEEN EACH JOIST. SEE DETAIL F.4

PARTY WALL AT LOAD BEARING WALLS

B. WALL FRAMING

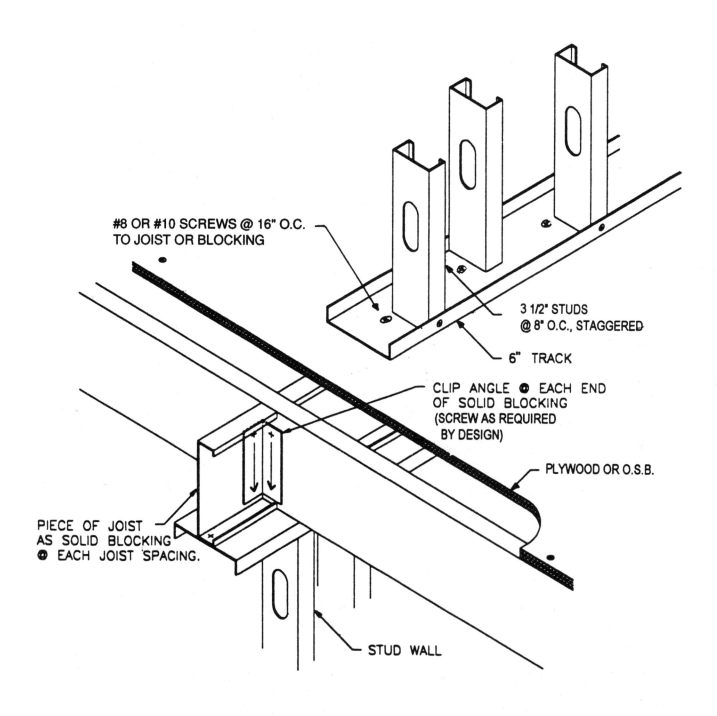

B.21 NON-LOAD BEARING SOUND PARTITION DETAILS

B. WALL FRAMING

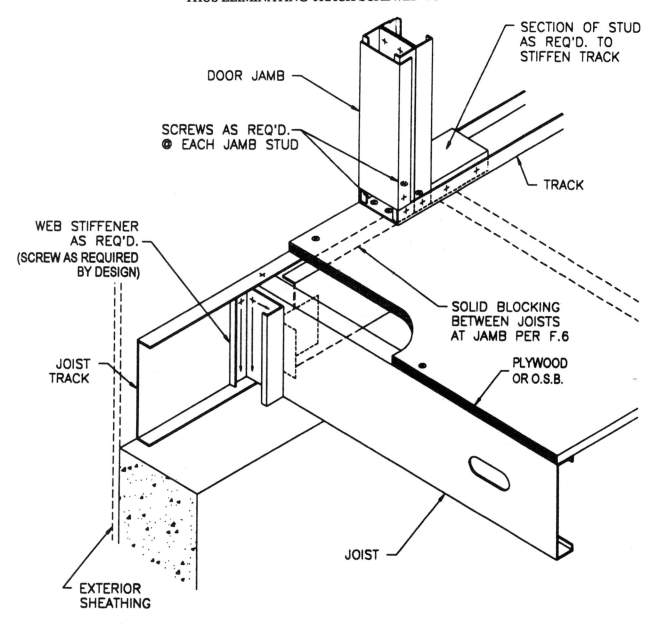

DOOR JAMB BASE AT FRAMING

B.22

B. WALL FRAMING

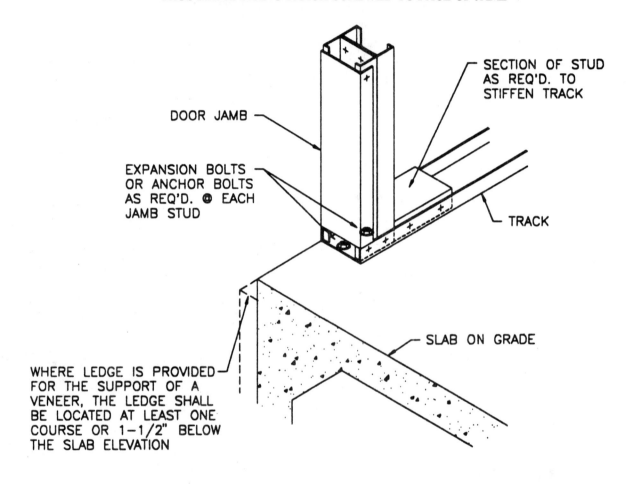

B.23 DOOR JAMB BASE AT SLAB ON GRADE

C. CEE CHANNEL ROOF SYSTEM

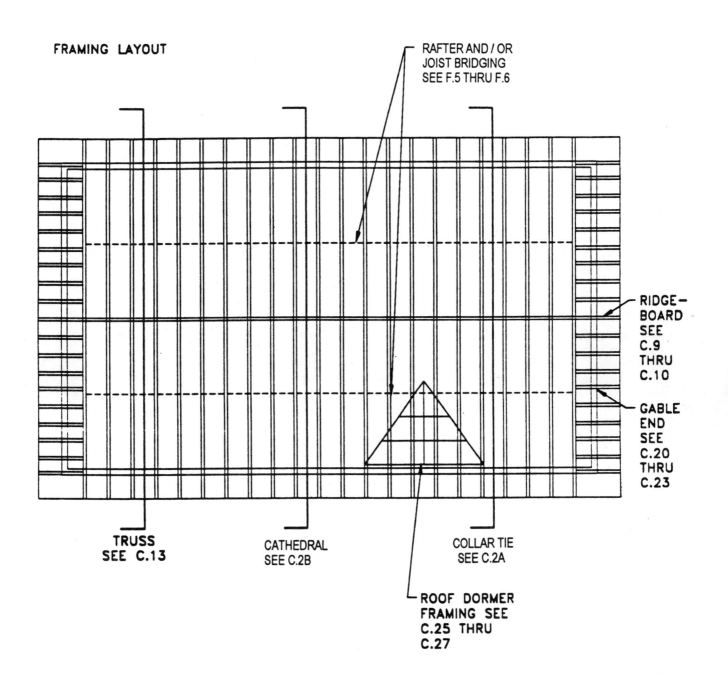

FRAMING LAYOUT

TYPICAL ROOF FRAMING PLAN

C.1

C. CEE CHANNEL ROOF SYSTEM

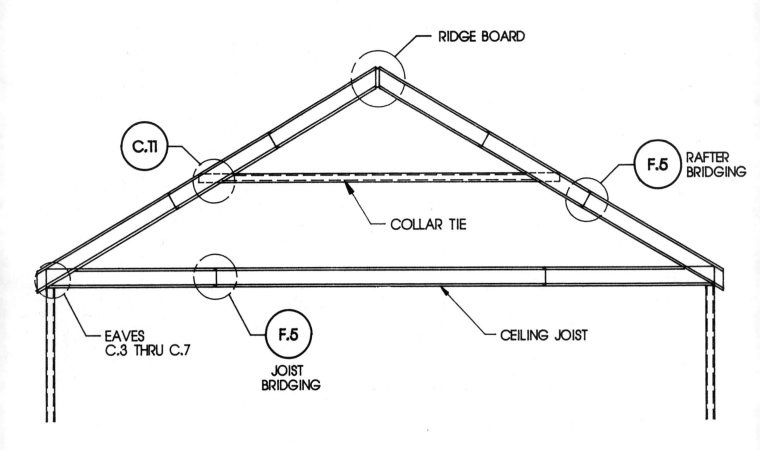

C.2a   TYPICAL RAFTER FRAMED ROOF SECTION

C. CEE CHANNEL ROOF SYSTEM

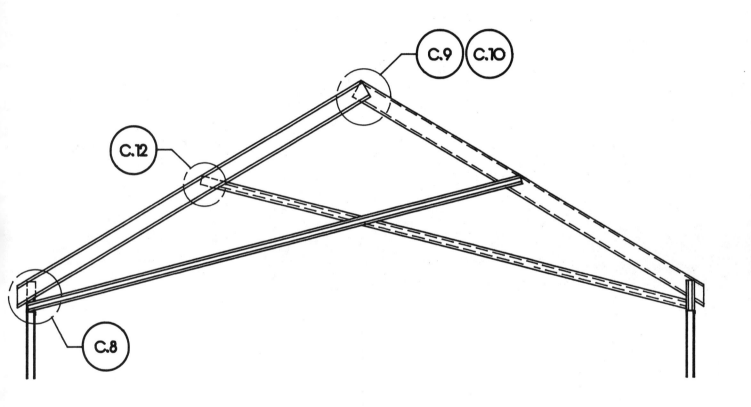

**TYPICAL RAFTER FRAMED
VAULTED / CATHEDRAL CEILING**

C.2b

C. CEE CHANNEL ROOF SYSTEM

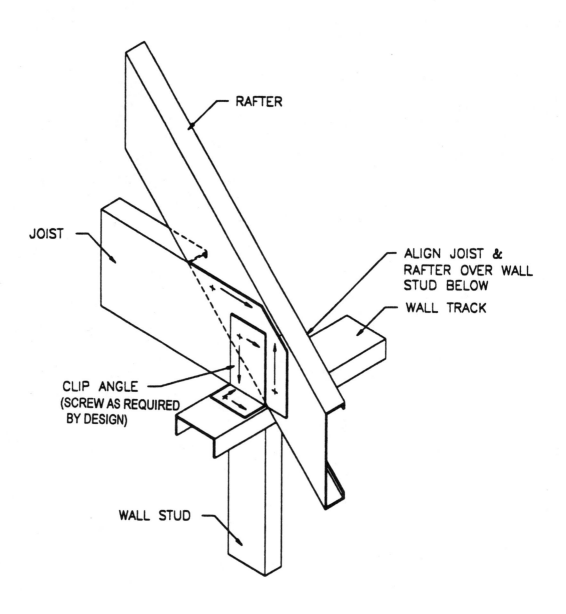

C.3   TRUSS EAVE DETAIL

C. CEE CHANNEL ROOF SYSTEM

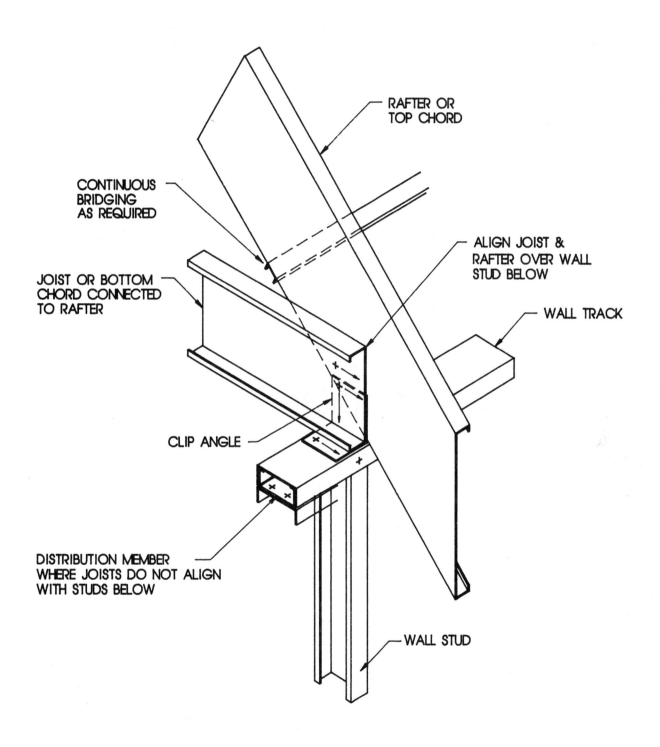

RAFTER EAVE DETAIL

C.4

C. CEE CHANNEL ROOF SYSTEM

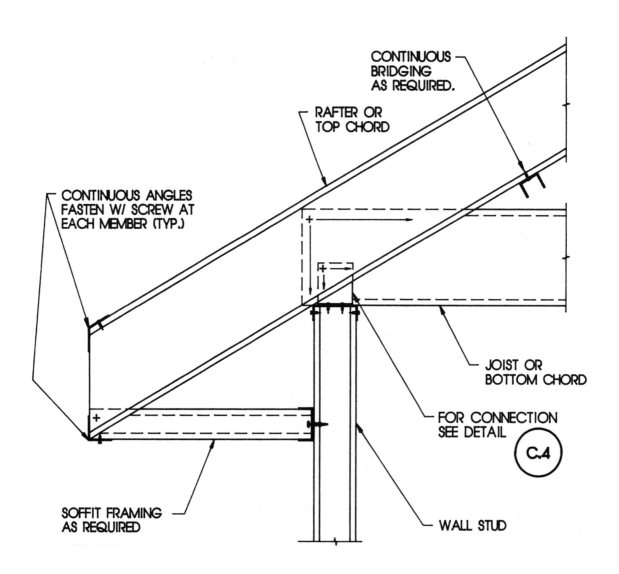

C.5   RAFTER EAVE DETAIL

C. CEE CHANNEL ROOF SYSTEM

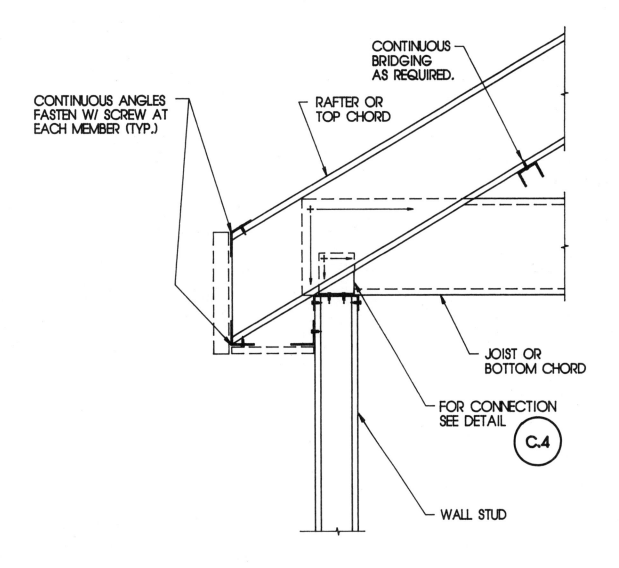

RAFTER EAVE DETAIL — C.6

C. CEE CHANNEL ROOF SYSTEM

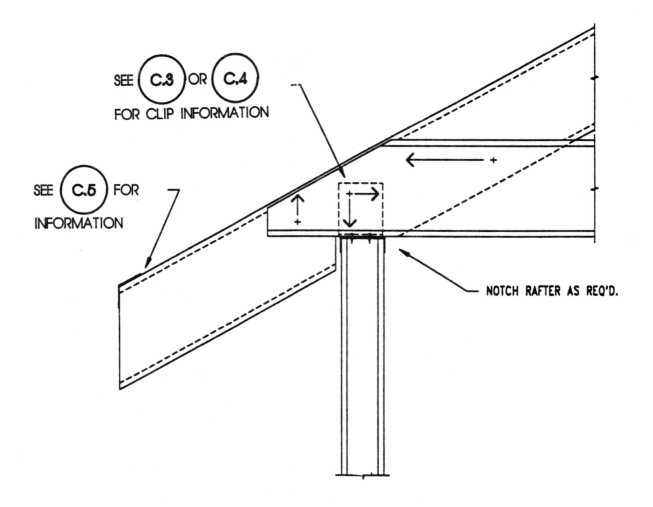

C.7 RAFTER EAVE DETAIL

C. CEE CHANNEL ROOF SYSTEM

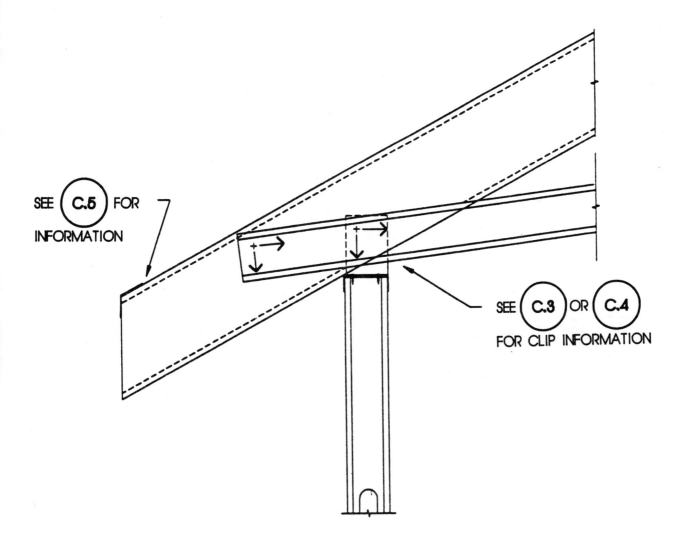

RAFTER EAVE DETAIL

C.8

C. CEE CHANNEL ROOF SYSTEM

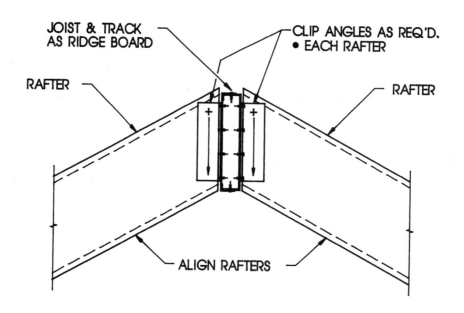

C.9   RIDGE BOARD DETAIL

C. CEE CHANNEL ROOF SYSTEM

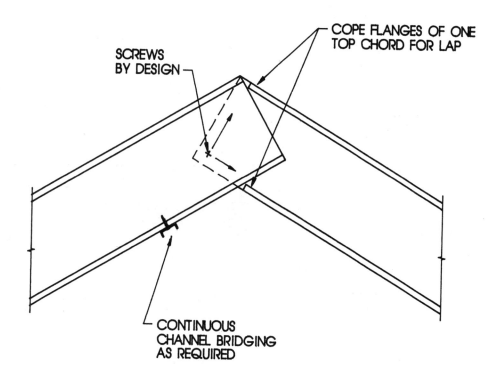

RIDGE DETAIL

C.10

C. CEE CHANNEL ROOF SYSTEM

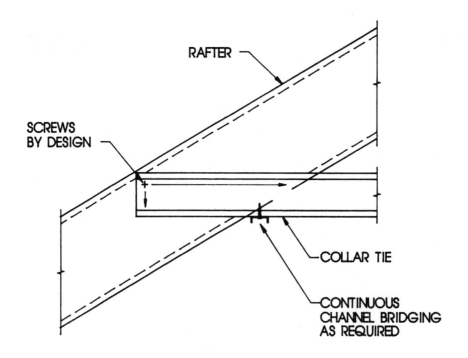

C.11   COLLAR TIE AT RAFTER DETAIL

C. CEE CHANNEL ROOF SYSTEM

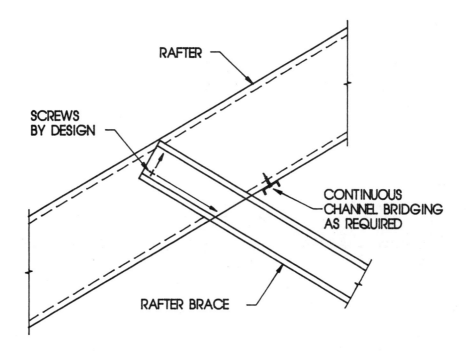

RAFTER TO DIAGONAL BRACE DETAIL   **C.12**

C. CEE CHANNEL ROOF SYSTEM

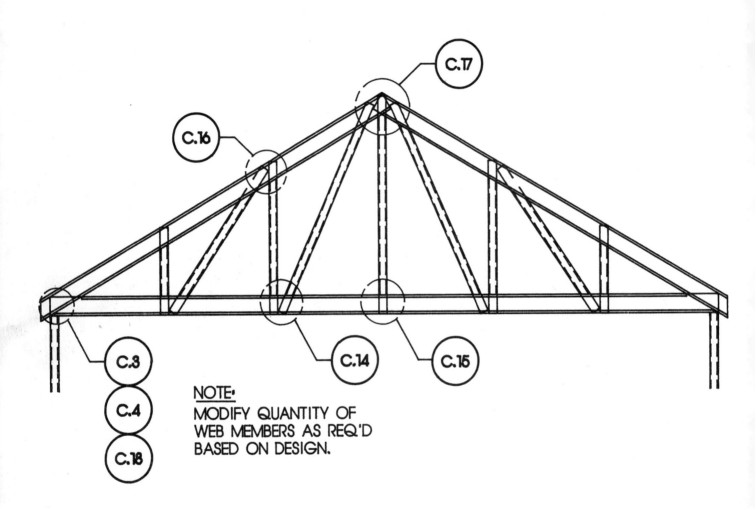

NOTE:
MODIFY QUANTITY OF WEB MEMBERS AS REQ'D BASED ON DESIGN.

C.13a   TYPICAL KING POST TRUSS PROFILE

C. CEE CHANNEL ROOF SYSTEM

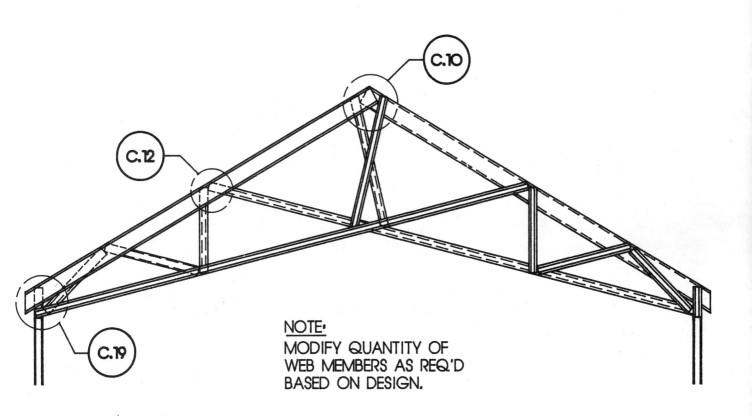

NOTE:
MODIFY QUANTITY OF
WEB MEMBERS AS REQ'D
BASED ON DESIGN.

TYPICAL SCISSORS TRUSS PROFILE   **C.13b**

C. CEE CHANNEL ROOF SYSTEM

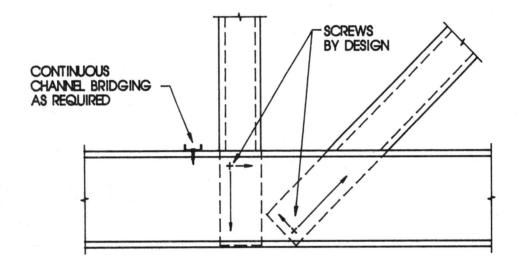

C.14   WEB TO BOTTOM CHORD DETAIL

C. CEE CHANNEL ROOF SYSTEM

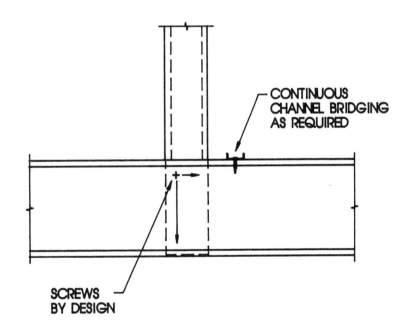

WEB TO BOTTOM CHORD DETAIL

C.15

C. CEE CHANNEL ROOF SYSTEM

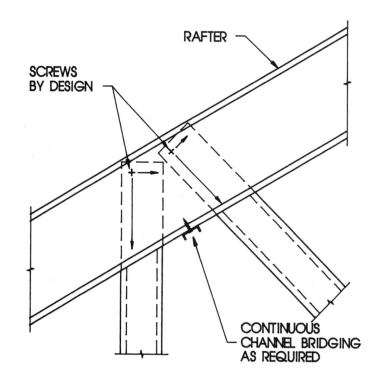

C.16   WEB TO TOP CHORD DETAIL

C. CEE CHANNEL ROOF SYSTEM

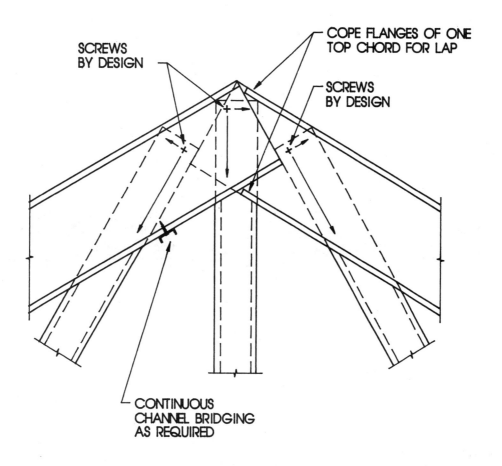

WEB AT PEAK OF TRUSS DETAIL

C.17

C. CEE CHANNEL ROOF SYSTEM

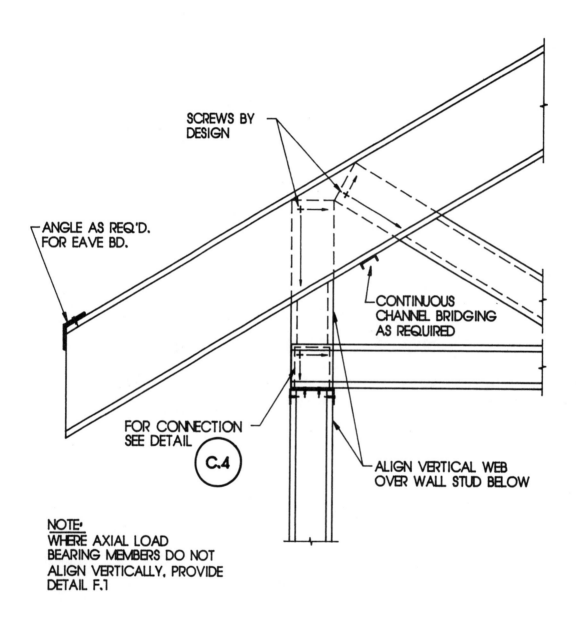

C.18 TRUSS END AT EXTERIOR WALL

C. CEE CHANNEL ROOF SYSTEM

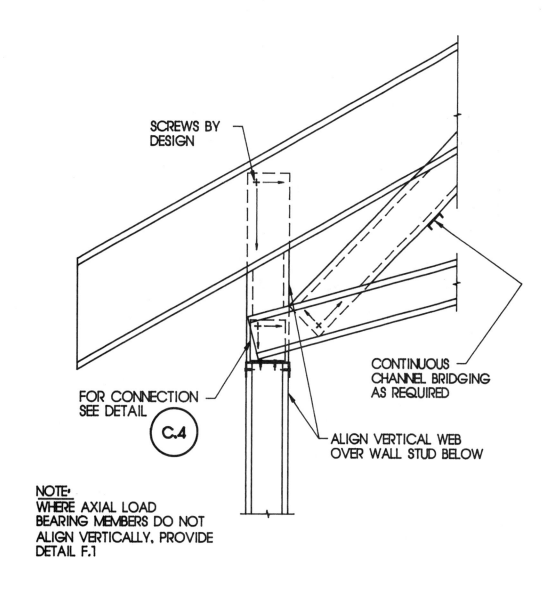

SCISSORS TRUSS END AT EXTERIOR WALL   C.19

C. CEE CHANNEL ROOF SYSTEM

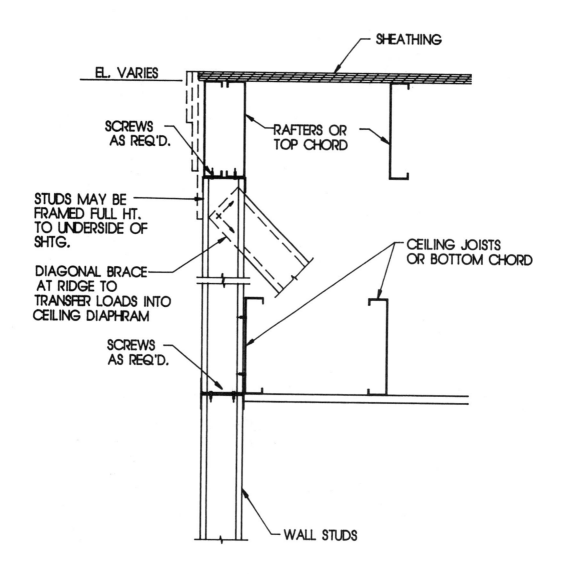

C.20   GABLE ROOF END DETAIL

C. CEE CHANNEL ROOF SYSTEM

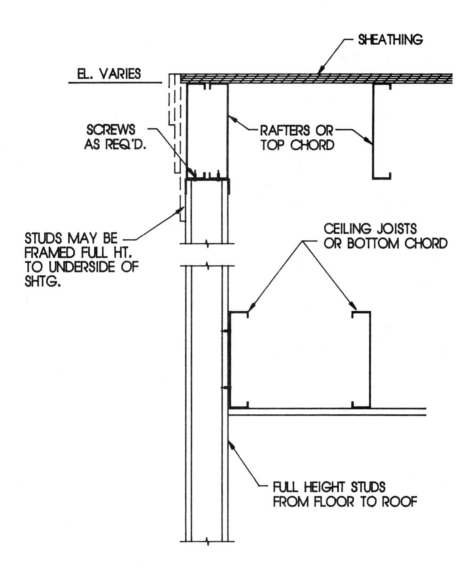

GABLE ROOF END DETAIL

C.21

C. CEE CHANNEL ROOF SYSTEM

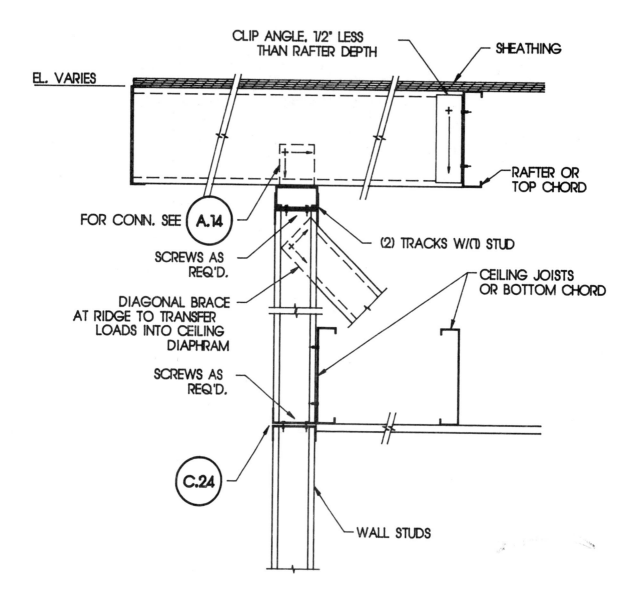

NOTES:
1. PROVIDE BRIDGING PER F.6 AT CEILING JOISTS AND ROOF RAFTERS.
2. PROVIDE CONTINUOUS BRIDGING BETWEEN RAFTERS AT WALL PER F.6

C.22 GABLE ROOF END DETAIL

C. CEE CHANNEL ROOF SYSTEM

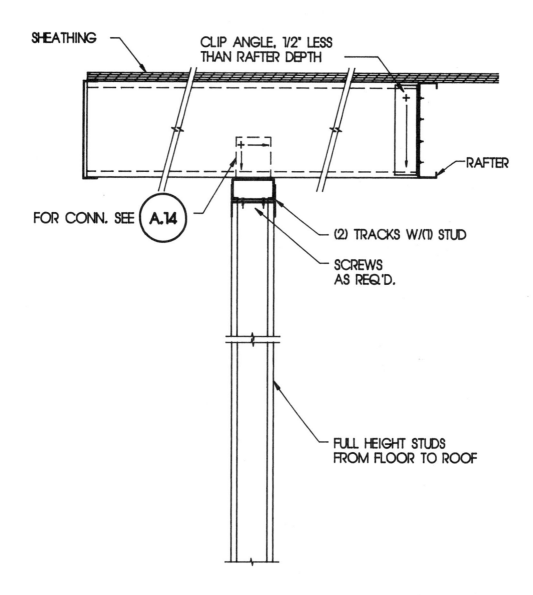

BALLOON FRAMED GABLE ROOF END DETAIL   **C.23**

C. CEE CHANNEL ROOF SYSTEM

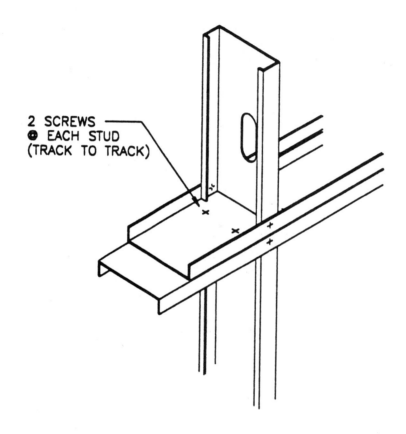

2 SCREWS @ EACH STUD (TRACK TO TRACK)

NOTE:
JOINT MUST BE BRACED DIAGONALLY OR HORIZONTALLY TO THE NEAREST ROOF OR FLOOR FRAMING MEMBER.

C.24 TRACK TO TRACK DETAIL

C. CEE CHANNEL ROOF SYSTEM

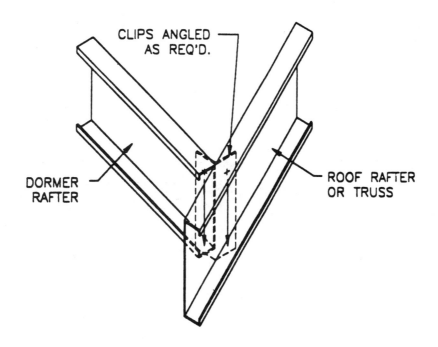

SUPPORTED MEMBER MAY BE CONNECTED BY CUTTING FLANGES - BENDING WEB TO DESIRED ANGLE & FASTENING DIRECTLY WITH SCREWS AS DESIGNED.

DORMER RAFTER AT ROOF RAFTER DETAIL   **C.25**

C. CEE CHANNEL ROOF SYSTEM

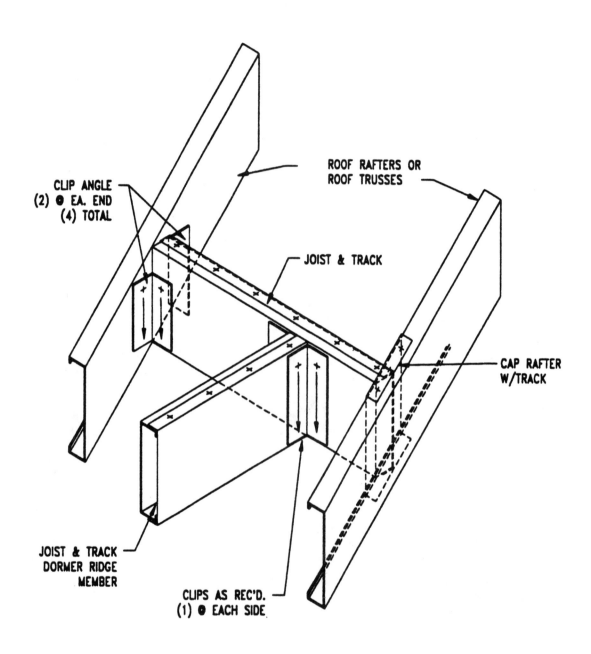

C.26   DORMER RIDGE AT MAIN ROOF DETAIL

C. CEE CHANNEL ROOF SYSTEM

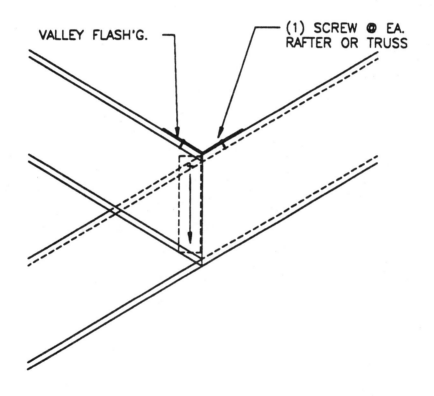

VALLEY FLASHING DETAIL　　C.27

C. CEE CHANNEL ROOF SYSTEM

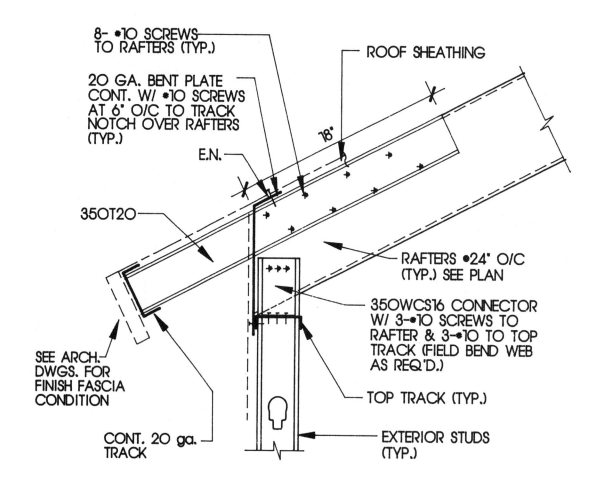

C.28 RAFTER WITH REDUCED SIZE OVERHANG

C. CEE CHANNEL ROOF SYSTEM

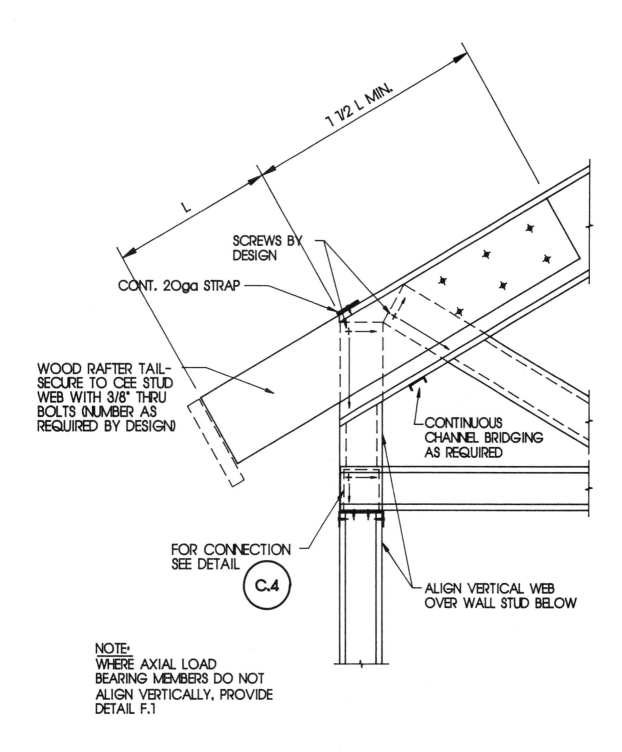

WOOD TAIL CONNECTION TO TRUSS

C.29

D. GUS TRUSS™ ROOF SYSTEM

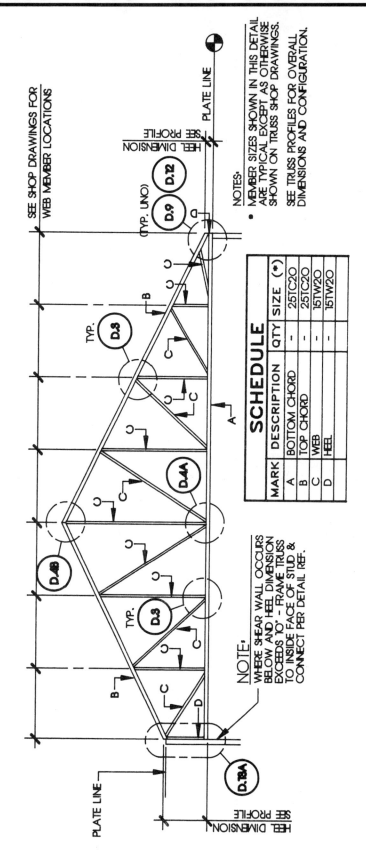

TYPICAL GUSS TRUSS™ ELEVATION

D.1a

D. GUS TRUSS™ ROOF SYSTEM

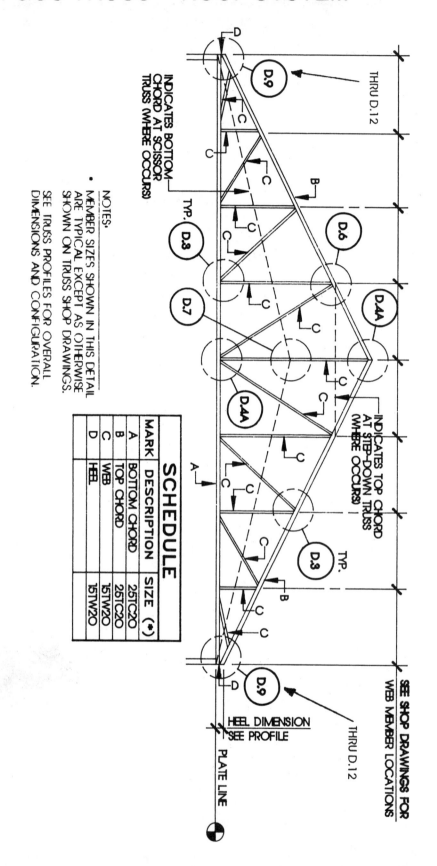

D.1b   TYPICAL GUSS TRUSS™ ELEVATION

D. GUS TRUSS™ ROOF SYSTEM

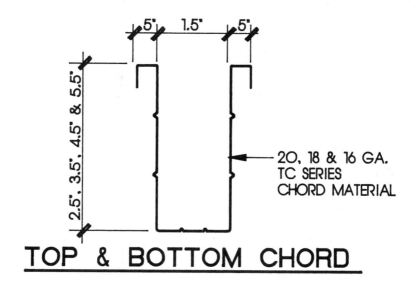

TOP & BOTTOM CHORD

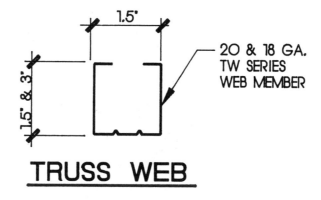

TRUSS WEB

GUS TRUSS™ MEMBER SECTIONS

D.2

D. GUS TRUSS™ ROOF SYSTEM

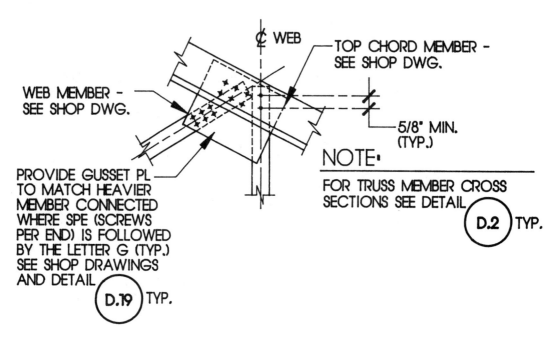

DETAIL AT TOP CHORD

NOTE: GUSSET PLATE MAY NOT BE REQUIRED IF CALCULATED NUMBER OF SCREWS CAN BE DIRECTLY APPLIED TO ALL JOINED WEBS THROUGH CHORD MEMBER.

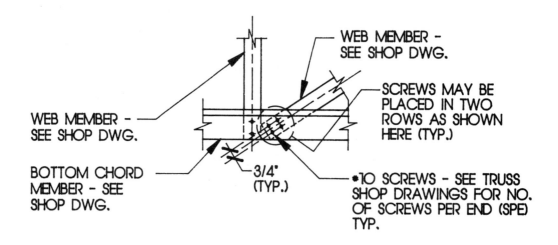

DETAIL AT BOTTOM CHORD

D.3   TRUSS WEB CONNECTION DETAIL

D. GUS TRUSS™ ROOF SYSTEM

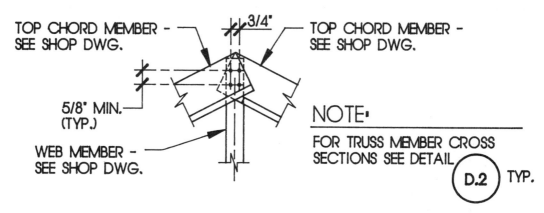

DETAIL AT TOP CHORD

NOTE: GUSSET PLATE MAY NOT BE REQUIRED IF CALCULATED NUMBER OF SCREWS CAN BE DIRECTLY APPLIED TO ALL JOINED WEBS THROUGH CHORD MEMBER.

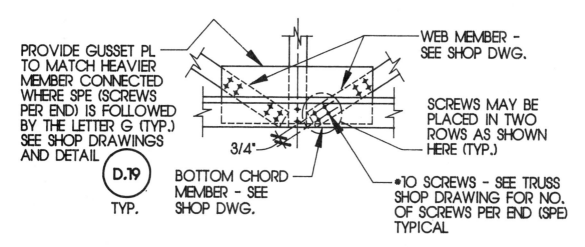

DETAIL AT BOTTOM CHORD

KING POST DETAIL D.4a

D. GUS TRUSS™ ROOF SYSTEM

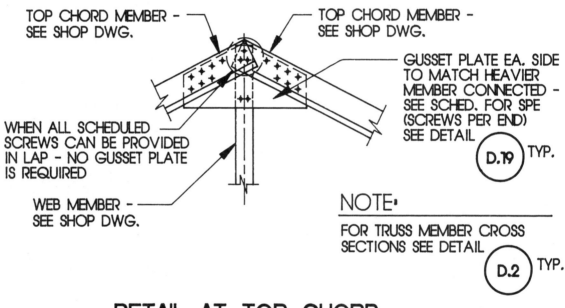

DETAIL AT TOP CHORD

NOTE: GUSSET PLATE MAY NOT BE REQUIRED IF CALCULATED NUMBER OF SCREWS CAN BE DIRECTLY APPLIED TO ALL JOINED WEBS THROUGH CHORD MEMBER.

D.4b   KING POST WITH GUSSET DETAIL

D. GUS TRUSS™ ROOF SYSTEM

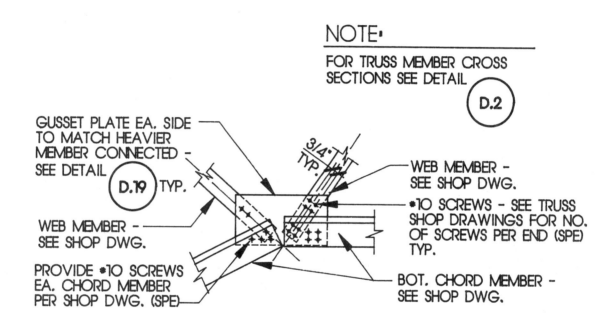

DETAIL AT BOTTOM CHORD

NOTE: GUSSET PLATE MAY NOT BE REQUIRED IF CALCULATED NUMBER OF SCREWS CAN BE DIRECTLY APPLIED TO ALL JOINED WEBS THROGH CHORD MEMBER.

SCISSORS TRUSS WITH CLIPPED CEILING DETAIL   D.5

D. GUS TRUSS™ ROOF SYSTEM

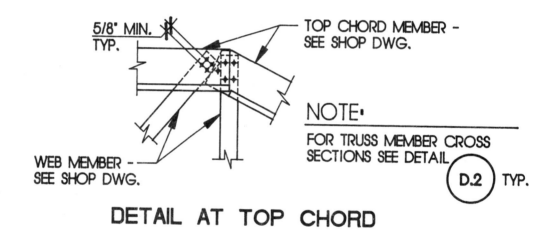

DETAIL AT TOP CHORD

NOTE: GUSSET PLATE MAY NOT BE REQUIRED IF CALCULATED NUMBER OF SCREWS CAN BE DIRECTLY APPLIED TO ALL JOINED WEBS THROUGH CHORD MEMBER.

D.6 STEP DOWN TOP CHORD DETAIL

D. GUS TRUSS™ ROOF SYSTEM

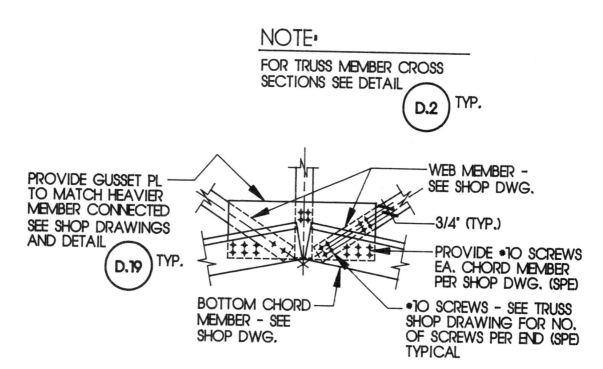

DETAIL AT BOTTOM CHORD

NOTE: GUSSET PLATE MAY NOT BE REQUIRED IF CALCULATED NUMBER OF SCREWS CAN BE DIRECTLY APPLIED TO ALL JOINED WEBS THROUGH CHORD MEMBER.

SCISSORS TRUSS BOTTOM CHORD DETAIL — D.7

D. GUS TRUSS™ ROOF SYSTEM

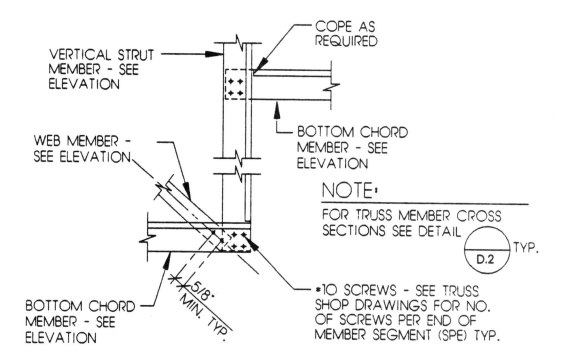

D.8   COMPOSITE TRUSS DETAIL (SCISSORS / COMMON)

D. GUS TRUSS™ ROOF SYSTEM

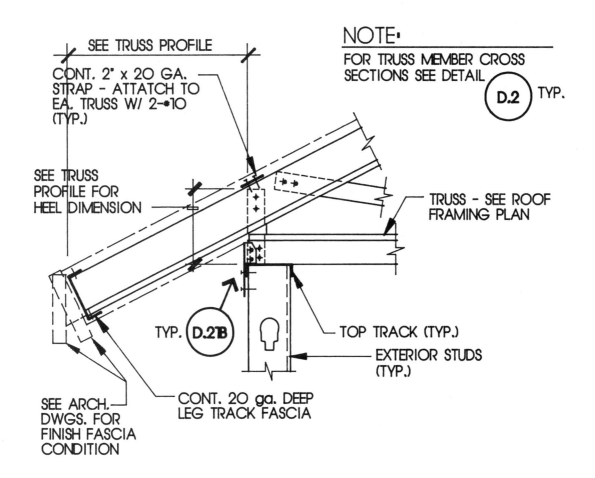

OVERHANG DETAIL – FLAT BOTTOM CHORD

D.9

D. GUS TRUSS™ ROOF SYSTEM

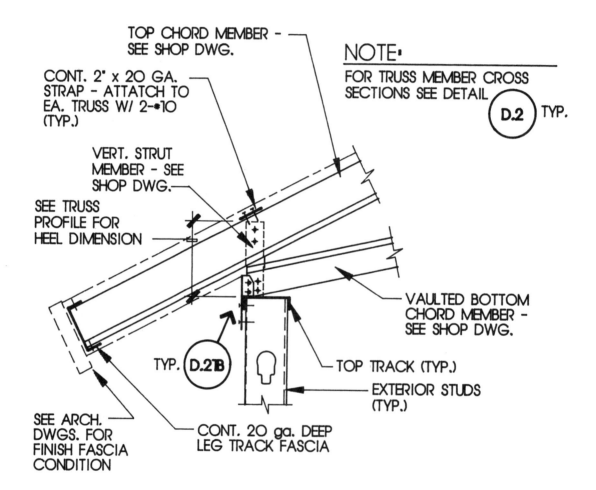

D.10 OVERHANG DETAIL – SCISSORS TRUSS

D. GUS TRUSS™ ROOF SYSTEM

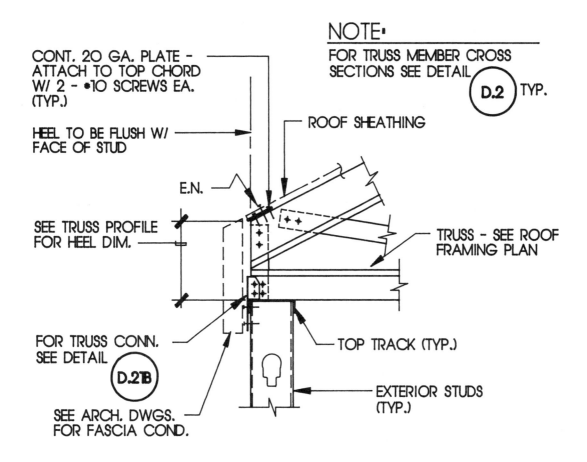

ZERO OVERHANG DETAIL – FLAT FASCIA

D.11

D. GUS TRUSS™ ROOF SYSTEM

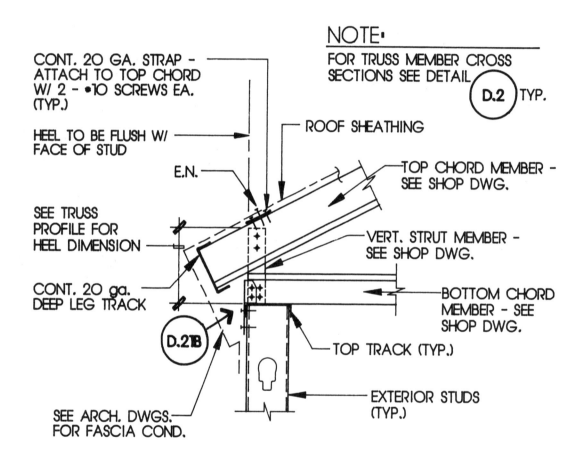

D.12   ZERO OVERHANG DETAIL – RAKED FASCIA

D. GUS TRUSS™ ROOF SYSTEM

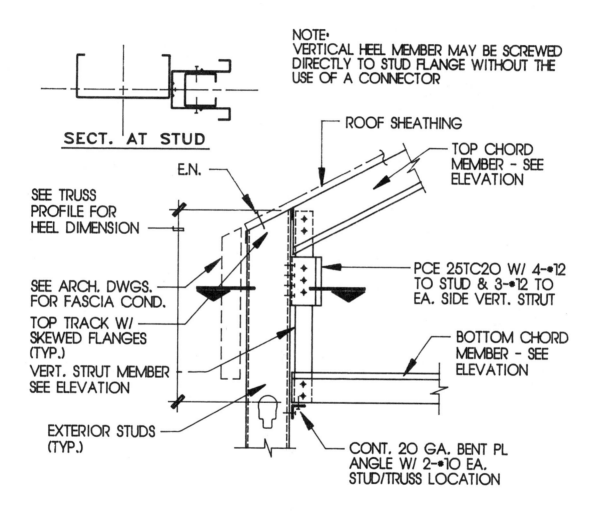

TRUSS CONNECTION TO FACE OF STUD — D.13a

D. GUS TRUSS™ ROOF SYSTEM

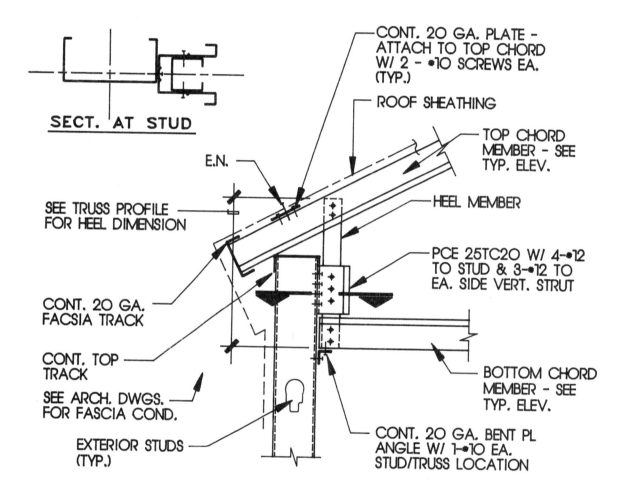

D.13b   TRUSS CONNECTION TO FACE OF STUD

D. GUS TRUSS™ ROOF SYSTEM

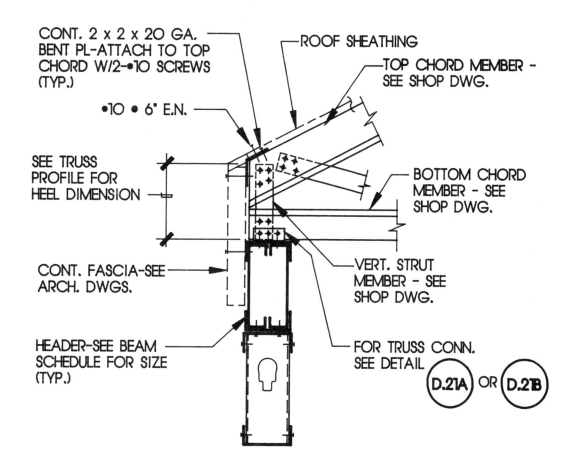

TRUSS CONNECTION TO HEADER – ZERO OVERHANG

D.14

D. GUS TRUSS™ ROOF SYSTEM

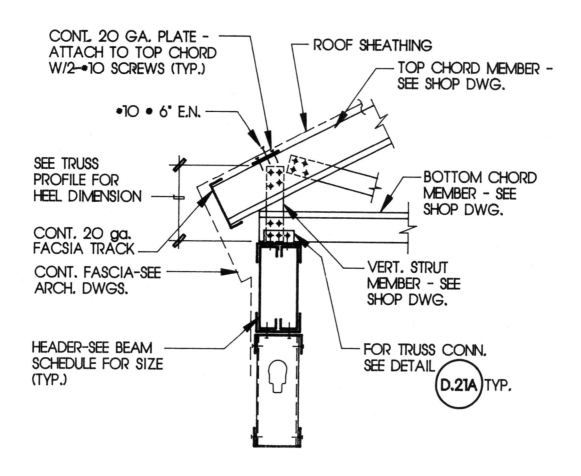

D.15   TRUSS CONNECTION TO HEADER – RAKED FASCIA

D. GUS TRUSS™ ROOF SYSTEM

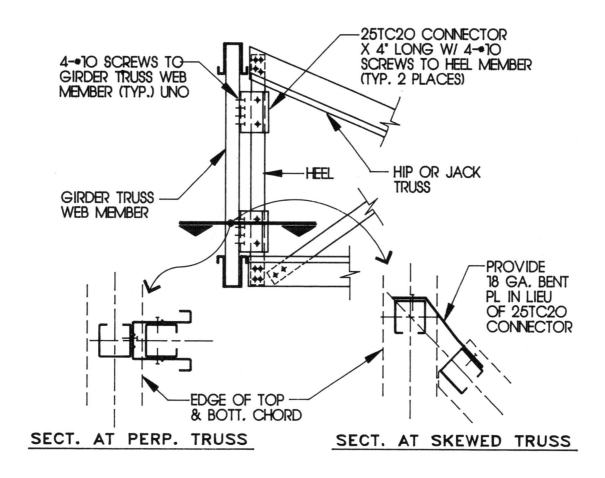

JACK TRUSS CONNECTION TO GIRDER TRUSS   **D.16**

D. GUS TRUSS™ ROOF SYSTEM

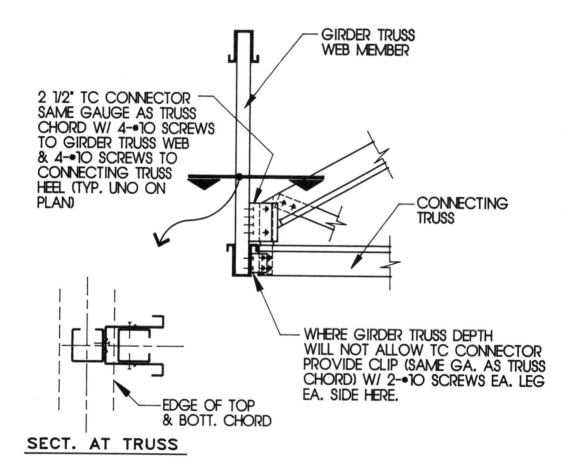

D.17 COMMON TRUSS CONNECTION TO GIRDER TRUSS

D. GUS TRUSS™ ROOF SYSTEM

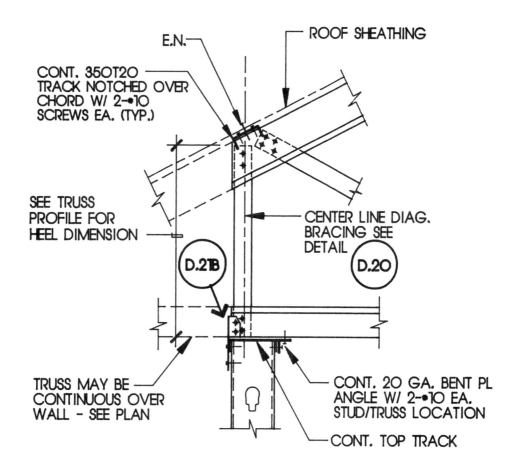

BLOCKED TRUSS HEEL SECTION

D.18

D. GUS TRUSS™ ROOF SYSTEM

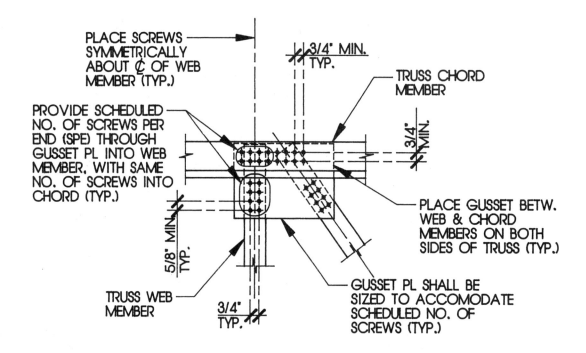

D.19 TYPICAL TRUSS GUSSET

D. GUS TRUSS™ ROOF SYSTEM

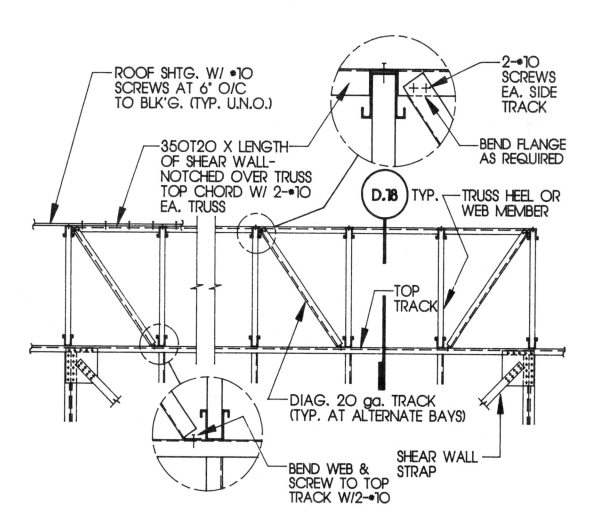

TRUSS BLOCKING DETAIL — D.20

D. GUS TRUSS™ ROOF SYSTEM

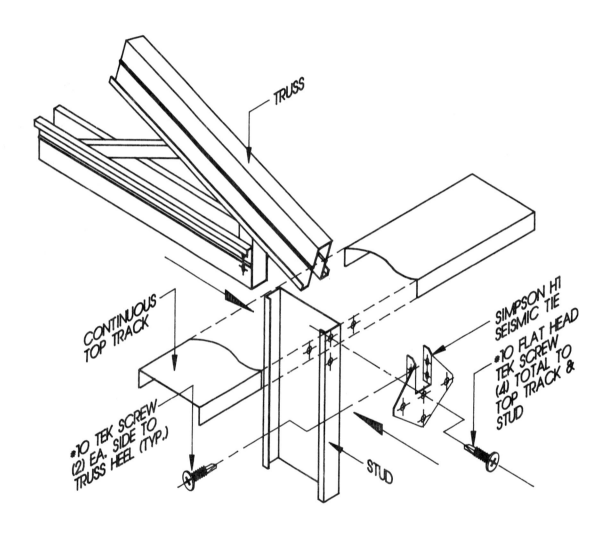

D.21 GUSS TRUSS™ TO TOP TRACK DETAIL

D. GUS TRUSS™ ROOF SYSTEM

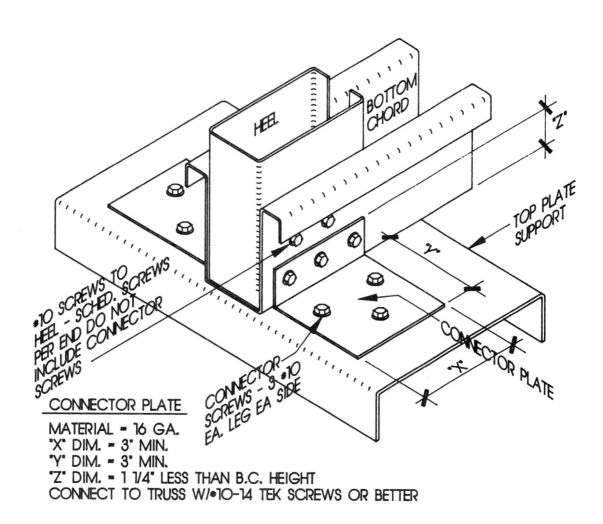

CONNECTOR PLATE

MATERIAL = 16 GA.
"X" DIM. = 3" MIN.
"Y" DIM. = 3" MIN.
"Z" DIM. = 1 1/4" LESS THAN B.C. HEIGHT
CONNECT TO TRUSS W/ #10-14 TEK SCREWS OR BETTER

BOTTOM CHORD TO TOP PLATE CONNECTION DETAIL — D.21a

D. GUS TRUSS™ ROOF SYSTEM

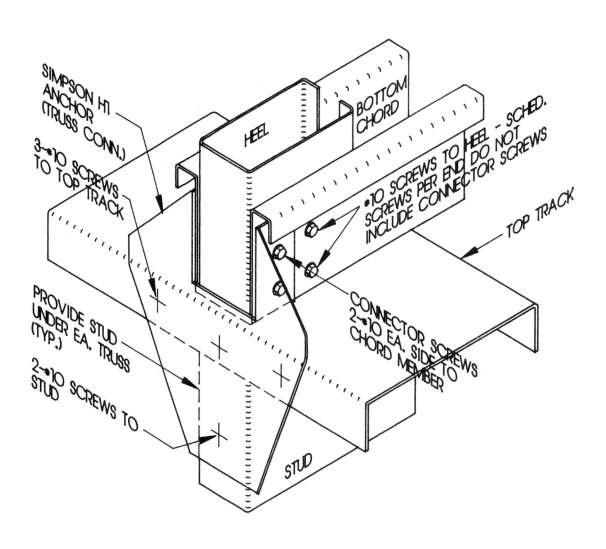

D.21b   BOTTOM CHORD TO TOP PLATE CONNECTION DETAIL

D. GUS TRUSS™ ROOF SYSTEM

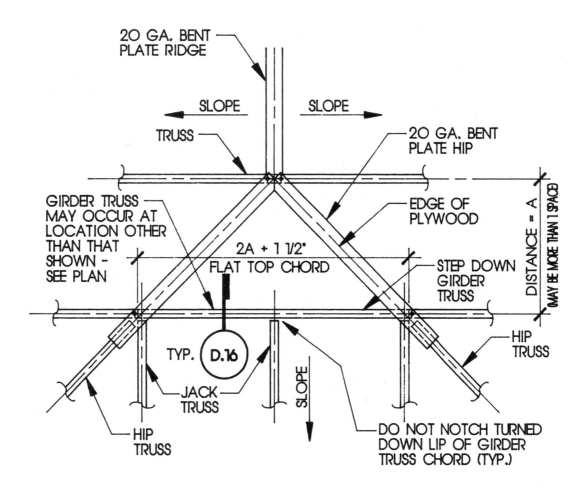

TYPICAL HIP ROOF PLAN — D.22

D. GUS TRUSS™ ROOF SYSTEM

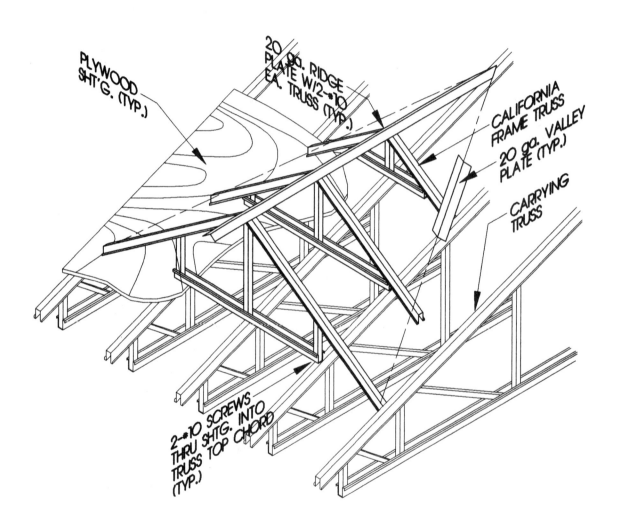

D.23 TYPICAL OVERFRAME
(CALIFORNIA FRAMING) TRUSSES

D. GUS TRUSS™ ROOF SYSTEM

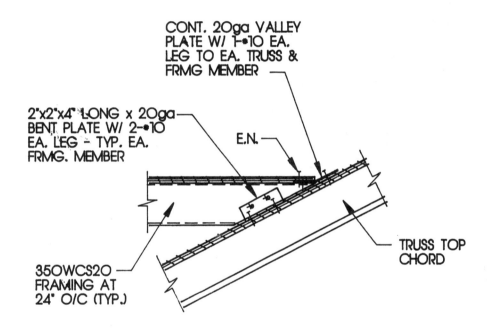

TYPICAL OVERFRAME DETAIL D.23a

D. GUS TRUSS™ ROOF SYSTEM

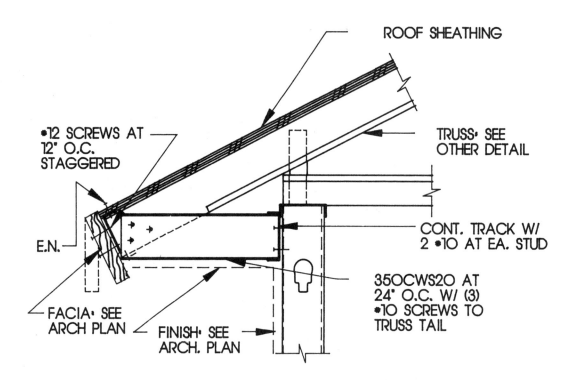

D.24   TYPICAL SOFFIT FRAMING DETAIL

D. GUS TRUSS™ ROOF SYSTEM

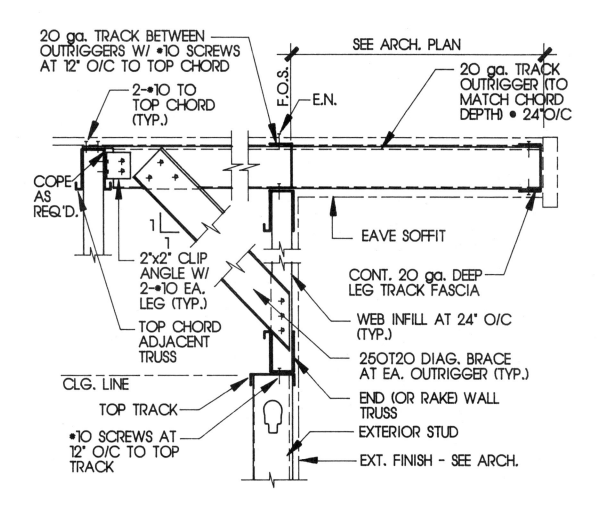

SECTION AT RAKE WALL TRUSS

D.25

D. GUS TRUSS™ ROOF SYSTEM

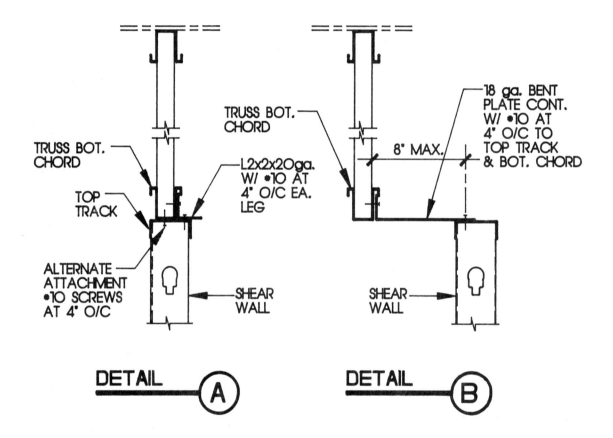

D.26   DRAG TRUSS TO SHEAR WALL TOP TRACK

D. GUS TRUSS™ ROOF SYSTEM

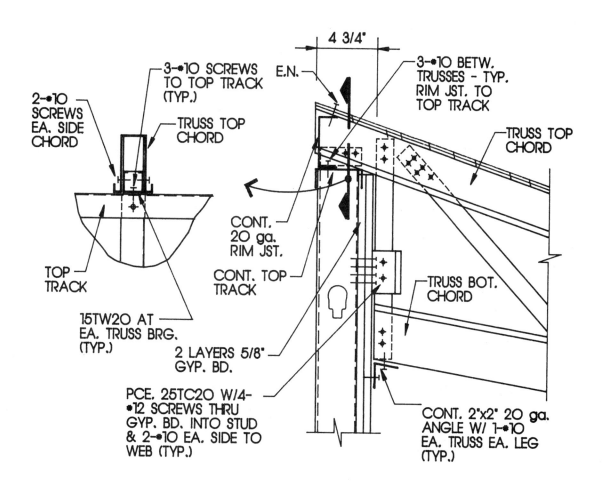

ROOF TRUSS CONNECTION AT PARTY WALL   **D.27**

D. GUS TRUSS™ ROOF SYSTEM

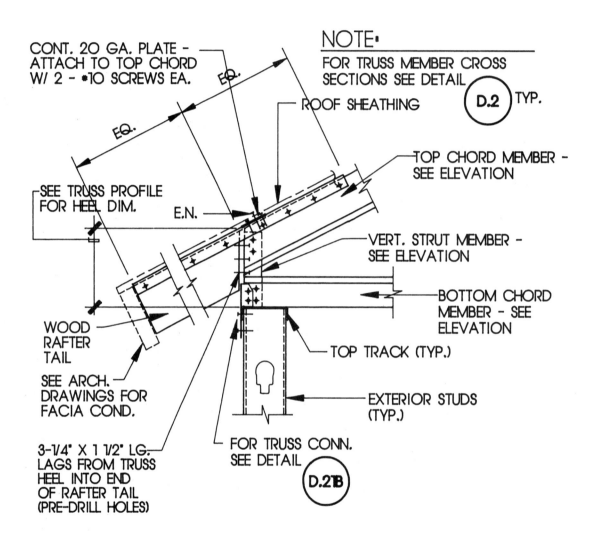

D.28   WOOD RAFTER TAIL CONNECTION TO GUS TRUSS™

D. GUS TRUSS™ ROOF SYSTEM

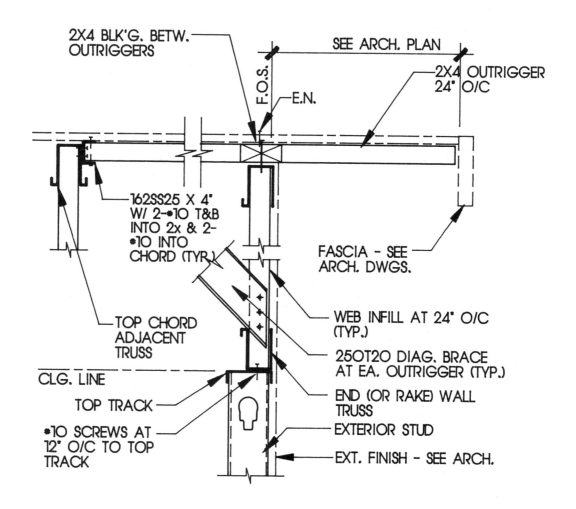

SECTION AT RAKE WALL TRUSS WITH WOOD OUTRIGGERS

D.29

D. GUS TRUSS™ ROOF SYSTEM

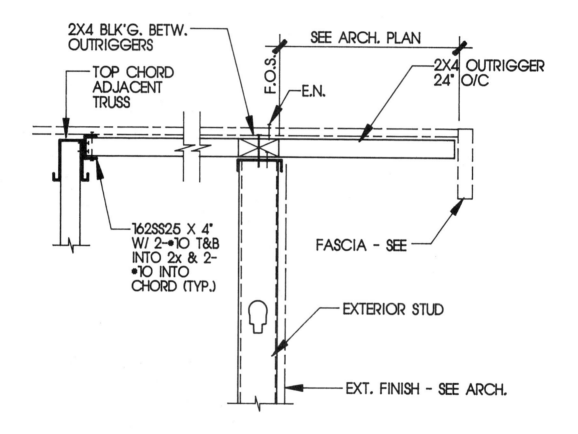

D.30   SECTION AT BALLON FRAMED RAKE WALL WITH WOOD OUTRIGGERS

D. GUS TRUSS™ ROOF SYSTEM

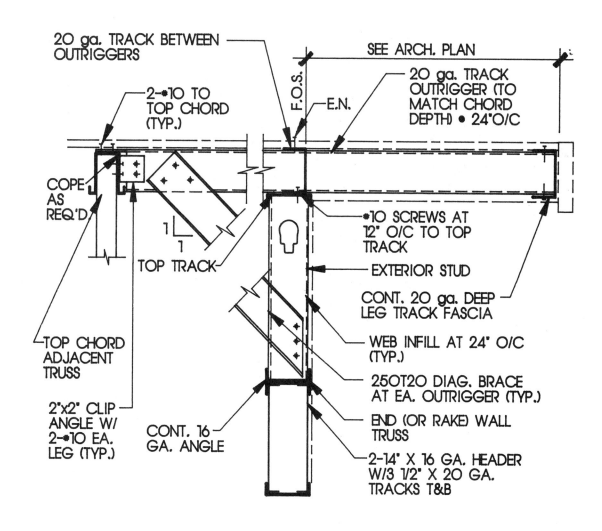

SECTION AT BALLOON FRAMED RAKE WALL OVER OPENING

D.31

D. GUS TRUSS™ ROOF SYSTEM

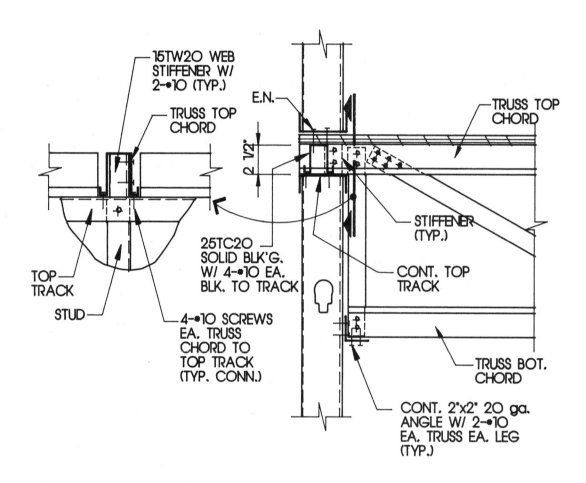

D.32a   TOP CHORD BEARING FLOOR TRUSS DETAIL

D. GUS TRUSS™ ROOF SYSTEM

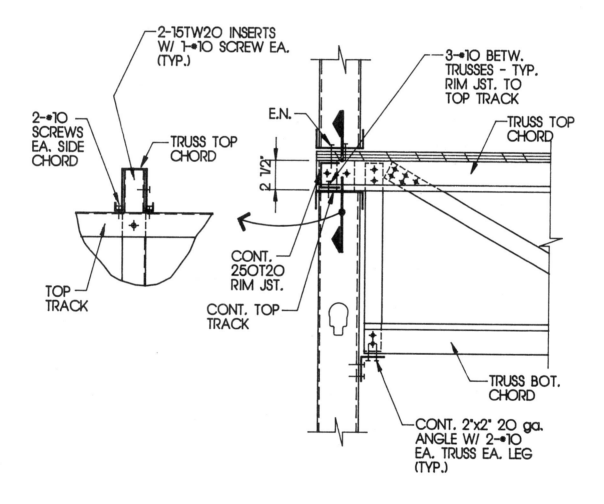

**TOP CHORD BEARING FLOOR TRUSS DETAIL
(ALTERNATE)**

D.32b

E. DETAILS AT EXTERIOR WALLS

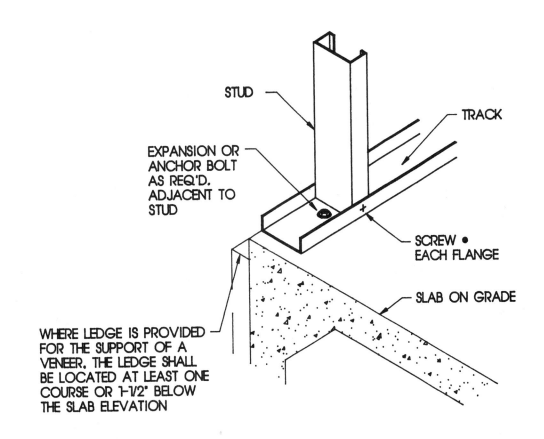

BASE OF WALL AT SLAB ON GRADE — E.1

E. DETAILS AT EXTERIOR WALLS

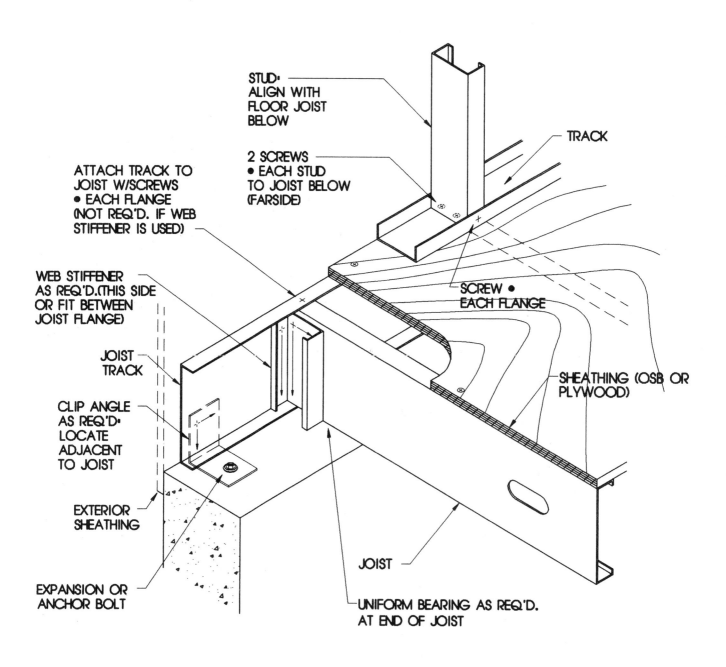

E.2 FLOOR JOISTS BEARING ON FOUNDATION

E. DETAILS AT EXTERIOR WALLS

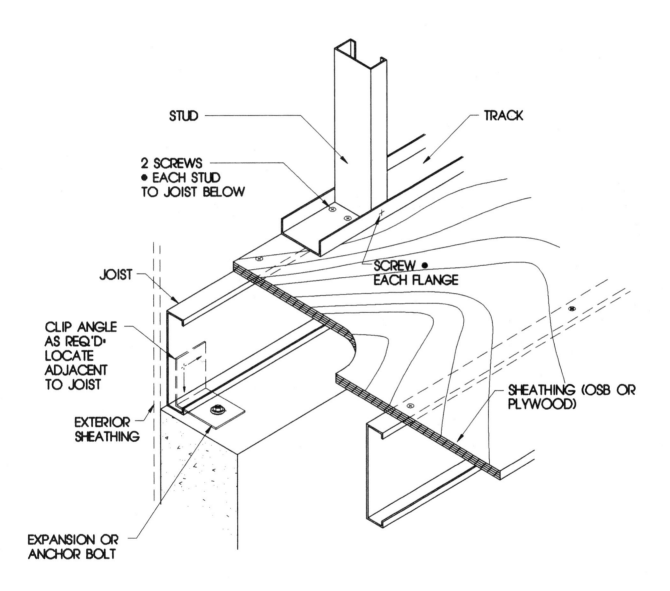

FLOOR JOISTS PARALLEL TO FOUNDATION   E.3

E. DETAILS AT EXTERIOR WALLS

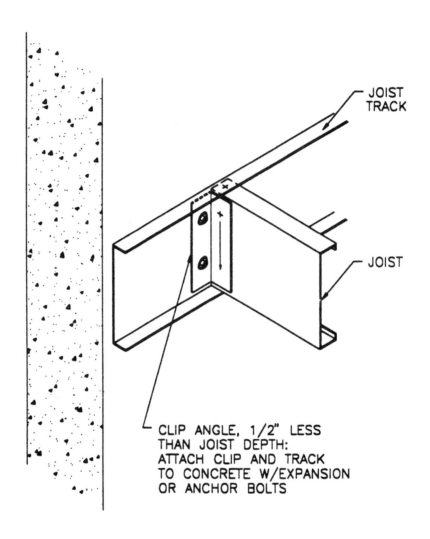

E.4 FLOOR JOIST SUPPORT AT CONTINUOUS WALL

E. DETAILS AT EXTERIOR WALLS

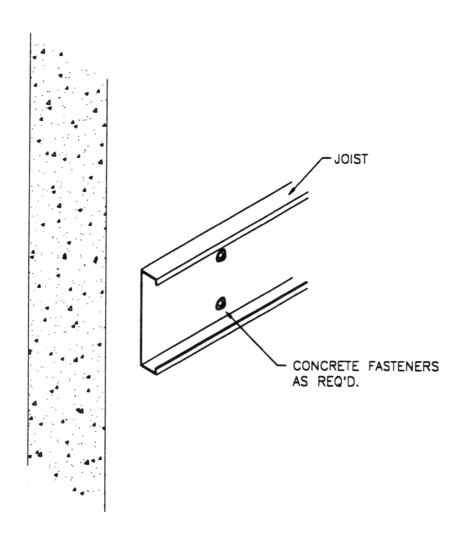

FLOOR JOISTS PARALLEL TO CONTINUOUS WALL   E.5

E. DETAILS AT EXTERIOR WALLS

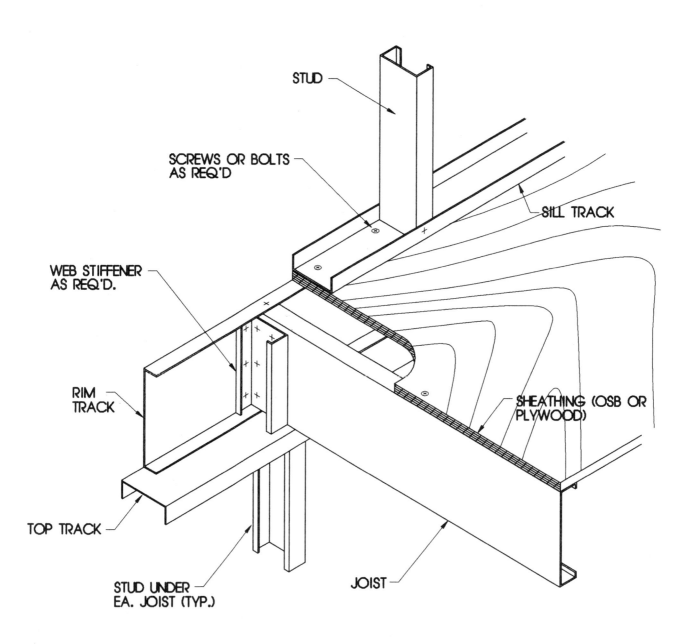

E.6   FLOOR FRAMING AT EXTERIOR WALL

E. DETAILS AT EXTERIOR WALLS

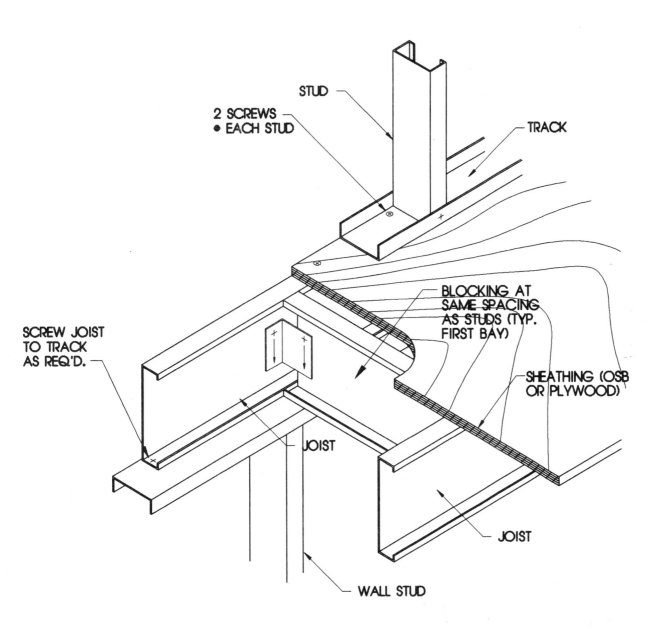

FLOOR JOISTS PARALLEL TO EXTERIOR WALL

E.7

E. DETAILS AT EXTERIOR WALLS

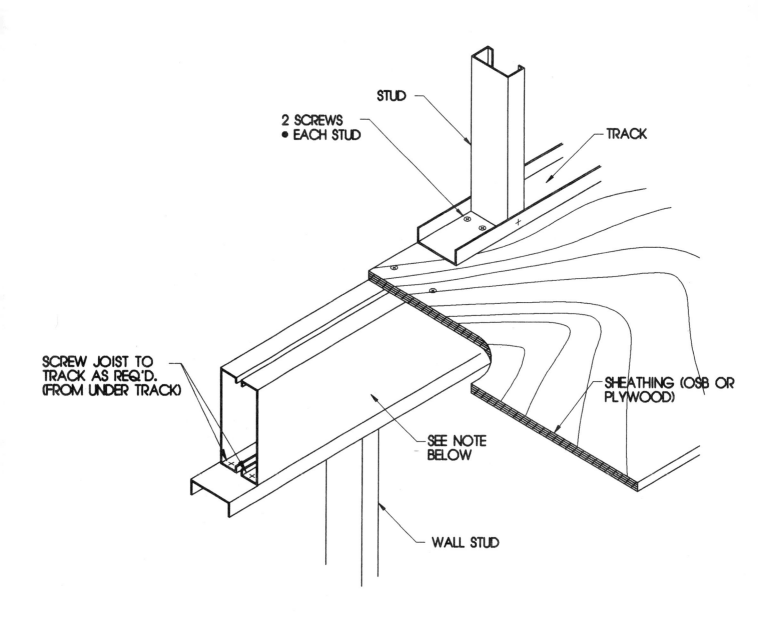

NOTE:
RIM JOIST MAY BE DOUBLED AS SHOWN AND MAY BE UTILIZED TO ELIMINATE THE NEED FOR ADDITIONAL DOOR OR WINDOW HEADERS. THIS DETAIL MAY ALSO BE USED WHERE FIRE RATED WALL CONSTRUCTION IS REQUIRED.

E.7a   FLOOR JOISTS PARALLEL TO EXTERIOR WALL (ALTERNATE)

E. DETAILS AT EXTERIOR WALLS

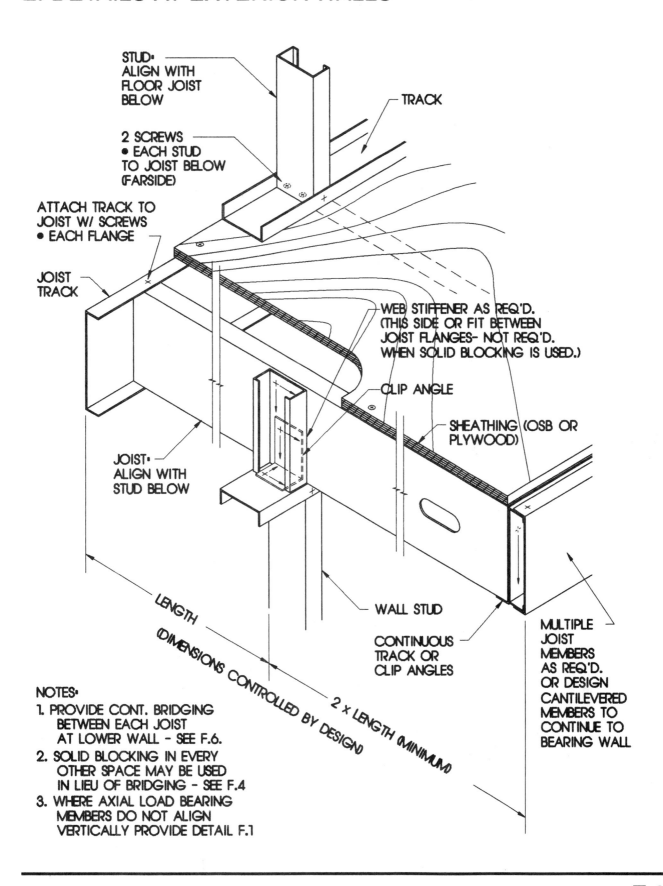

FLOOR CANTILEVER

E.8

E. DETAILS AT EXTERIOR WALLS

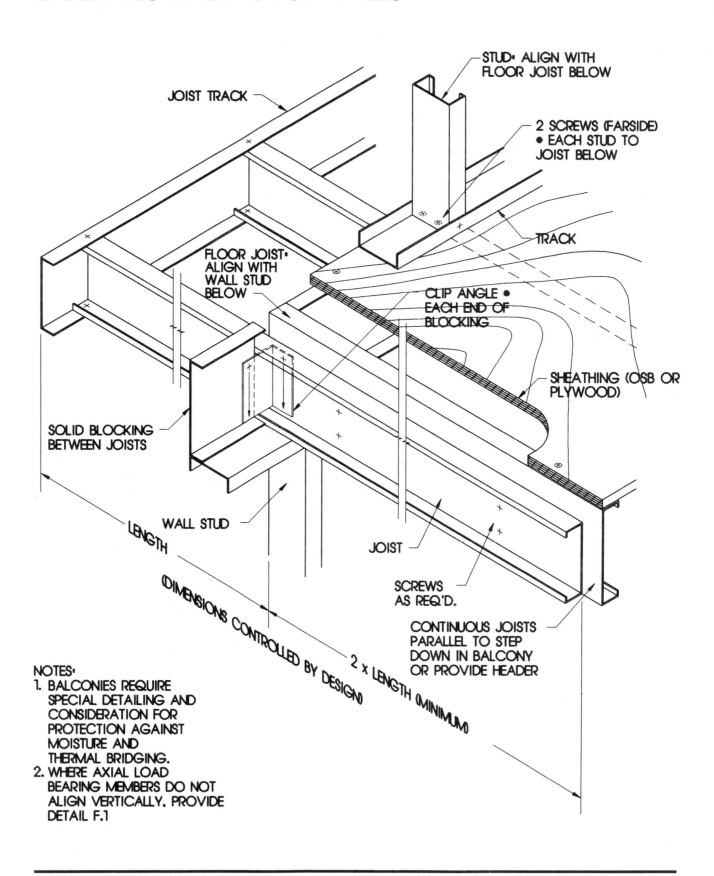

NOTES:
1. BALCONIES REQUIRE SPECIAL DETAILING AND CONSIDERATION FOR PROTECTION AGAINST MOISTURE AND THERMAL BRIDGING.
2. WHERE AXIAL LOAD BEARING MEMBERS DO NOT ALIGN VERTICALLY, PROVIDE DETAIL F.1

E.9　BALCONY WITH STEP DOWN

E. DETAILS AT EXTERIOR WALLS

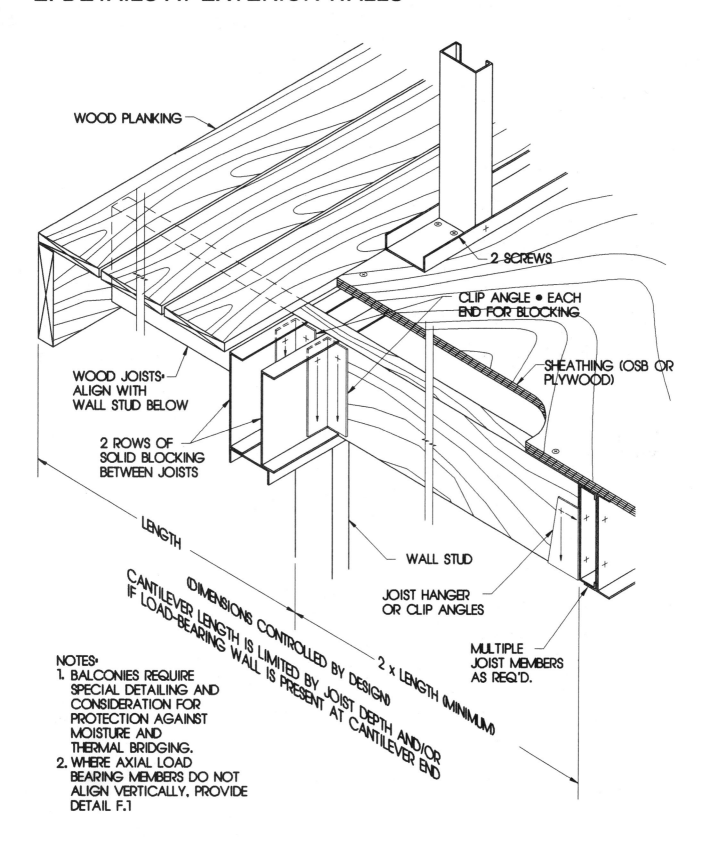

WOOD DECK BALCONY

E.10

E. DETAILS AT EXTERIOR WALLS

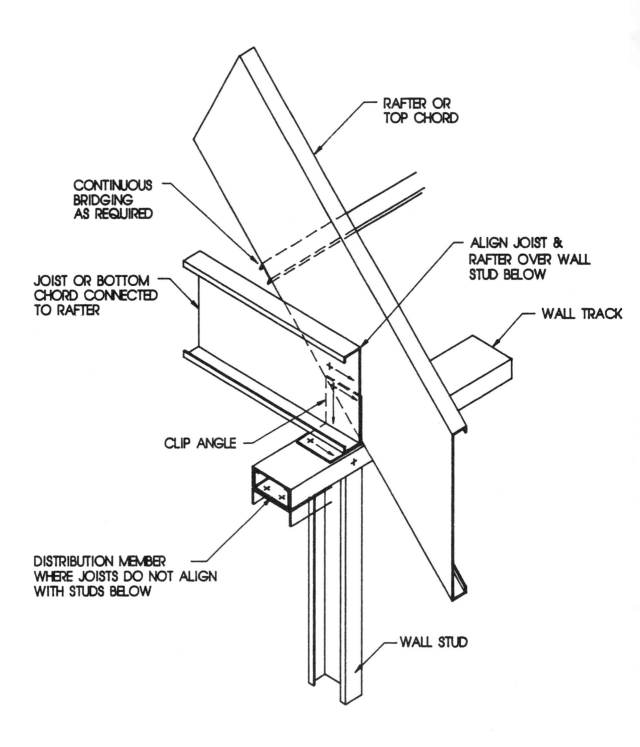

E.11   ROOF EAVE

E. DETAILS AT EXTERIOR WALLS

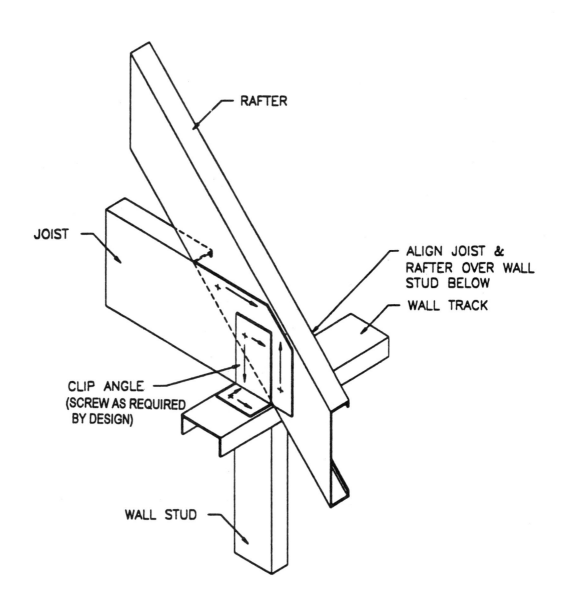

ROOF TRUSS EAVE   **E.12**

E. DETAILS AT EXTERIOR WALLS

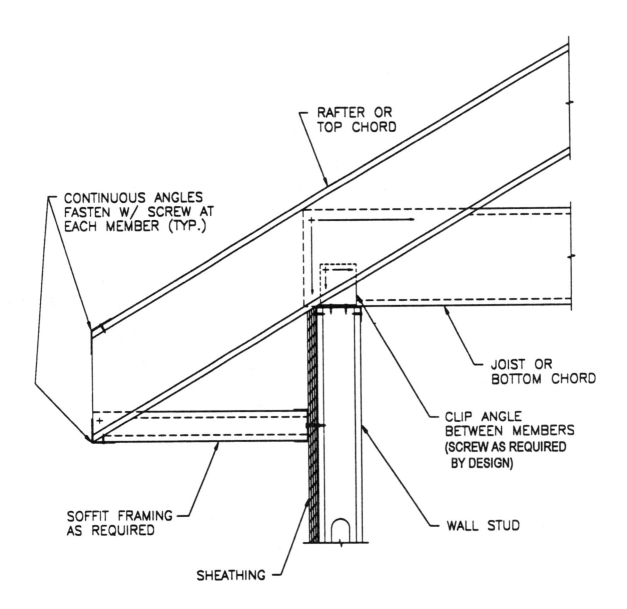

E.13   ROOF EAVE AND SOFFIT

E. DETAILS AT EXTERIOR WALLS

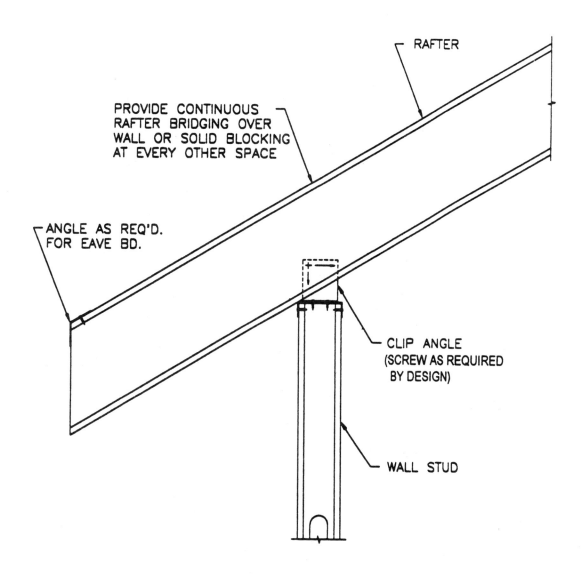

ROOF EAVE AT CATHEDRAL CEILING — E.14

E. DETAILS AT EXTERIOR WALLS

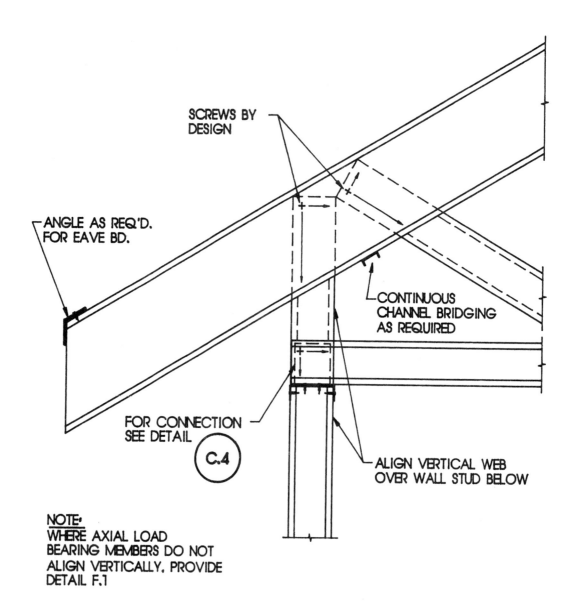

E.15   ROOF TRUSS BEARING

E. DETAILS AT EXTERIOR WALLS

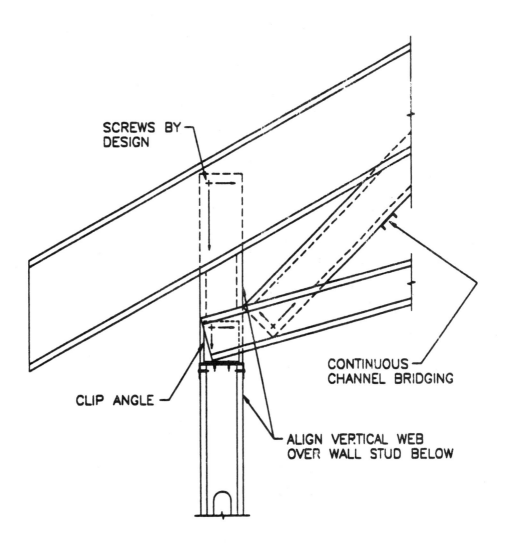

ROOF SCISSORS TRUSS BEARING   E.16

E. DETAILS AT EXTERIOR WALLS

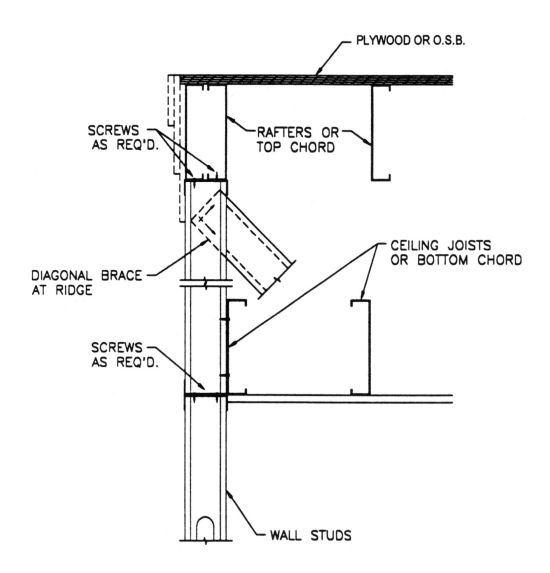

NOTE:
PROVIDE BRIDGING PER F.6
AT CEILING JOISTS AND RAFTERS.

E.17   ROOF GABLE END

E. DETAILS AT EXTERIOR WALLS

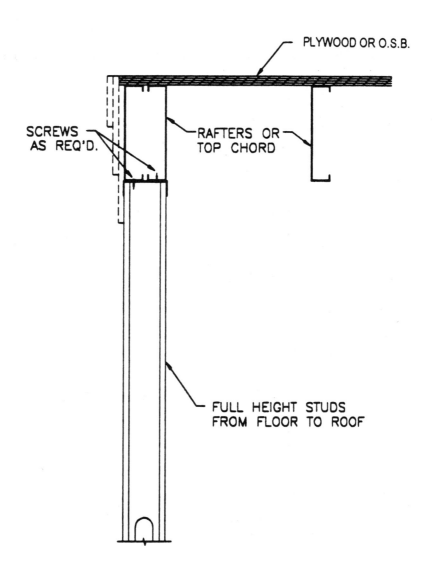

ROOF GABLE END AT CATHEDRAL   **E.18**

E. DETAILS AT EXTERIOR WALLS

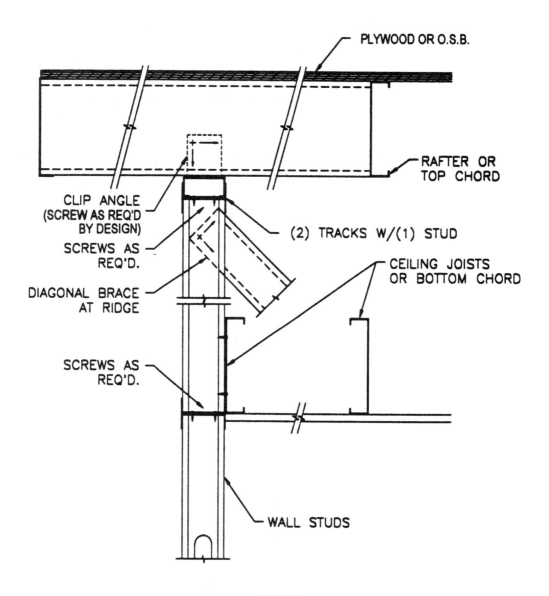

E.19 CANTILEVERED ROOF GABLE END

E. DETAILS AT EXTERIOR WALLS

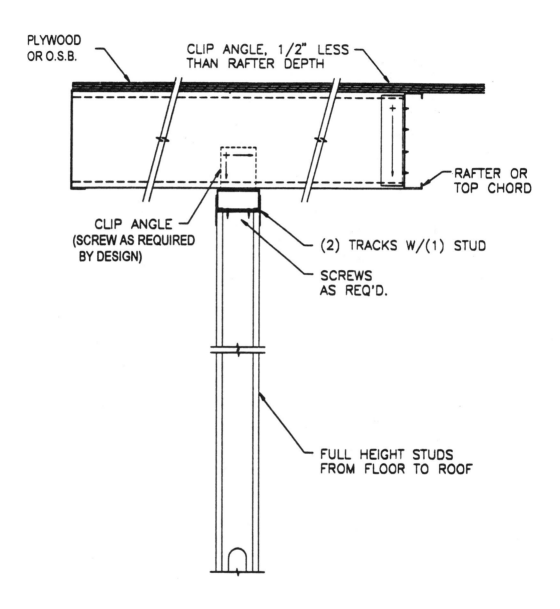

CANTILEVERED GABLE END AT CATHEDRAL — E.20

E. DETAILS AT EXTERIOR WALLS

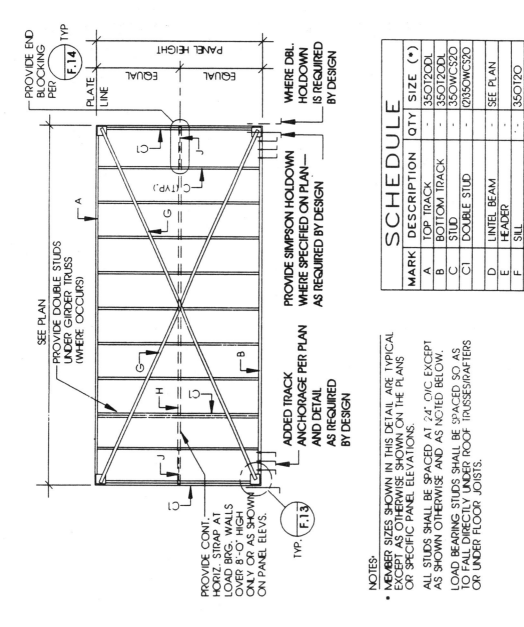

E.21a   TYPICAL SHEAR PANEL ELEVATION

E. DETAILS AT EXTERIOR WALLS

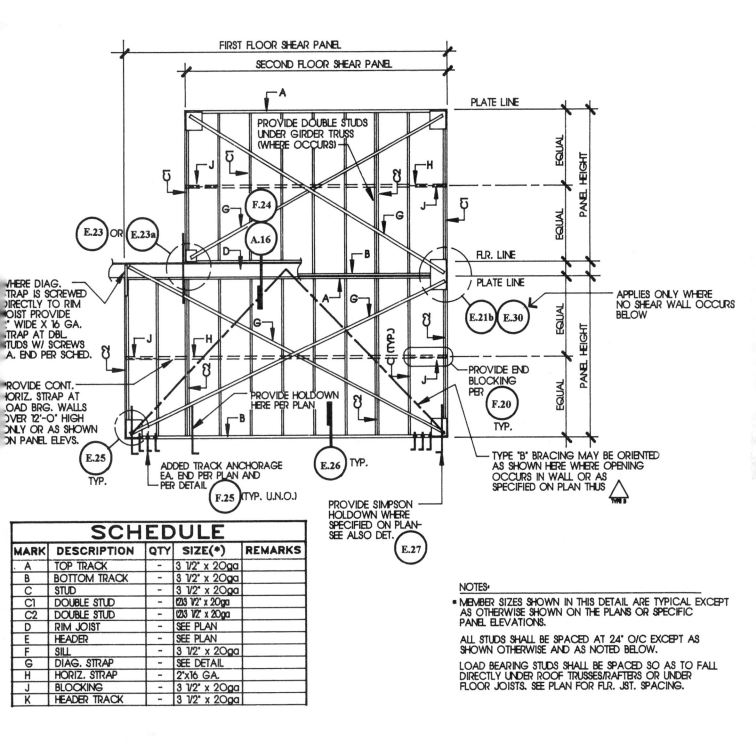

TYPICAL SHEAR PANEL ELEVATION – 2 STORY

E.21b

E. DETAILS AT EXTERIOR WALLS

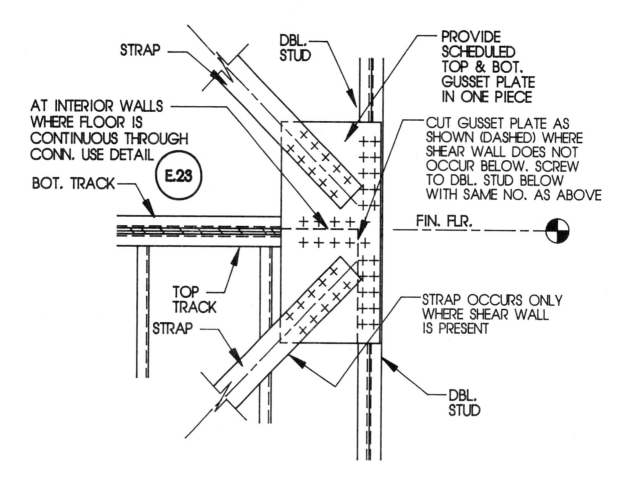

E.22   STACKED SECOND FLOOR SHEAR WALL WITH COMMON GUSSET PLATE

E. DETAILS AT EXTERIOR WALLS

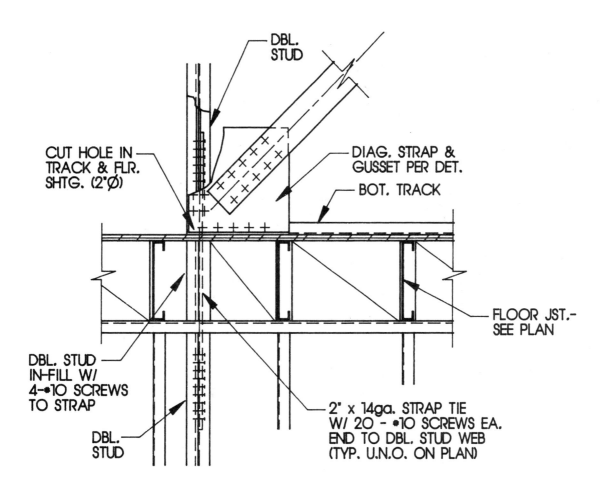

SECOND FLOOR SHEAR WALL STRAP TIE HOLDOWN DETAIL

E.23

E. DETAILS AT EXTERIOR WALLS

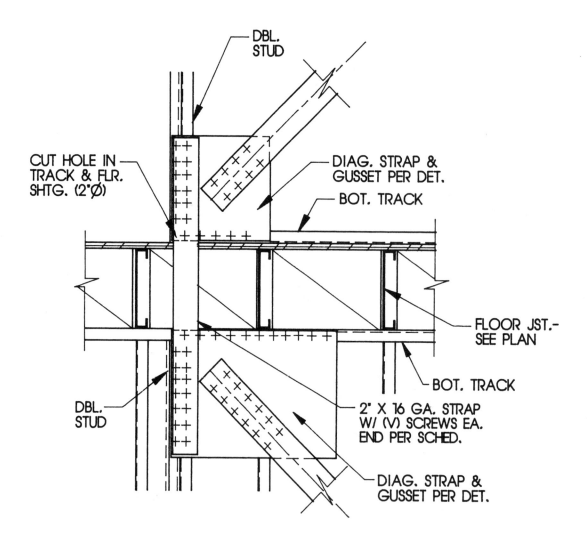

E.23a   SECOND FLOOR SHEAR WALL STRAP TIE HOLDOWN DETAIL (ALTERNATE)

E. DETAILS AT EXTERIOR WALLS

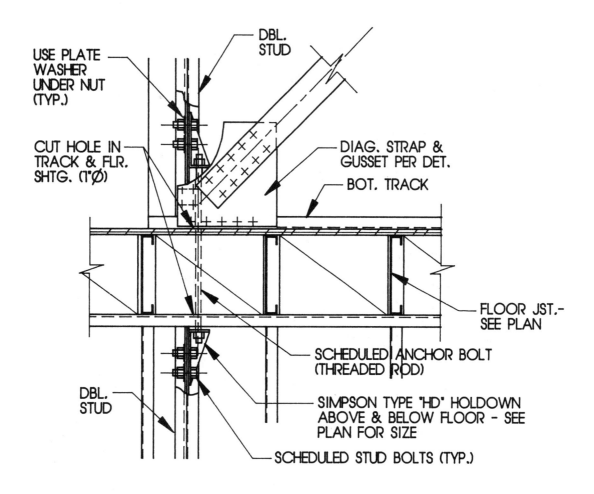

SECOND FLOOR SHEAR WALL HOLDOWN DETAIL — E.24

E. DETAILS AT EXTERIOR WALLS

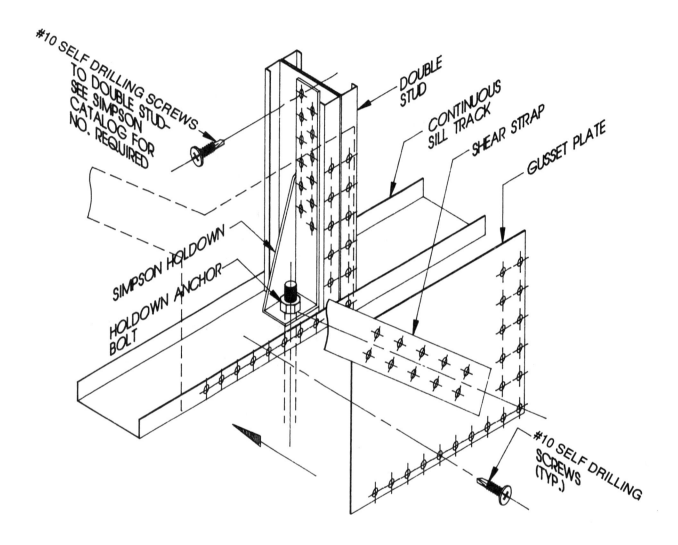

E.25   SHEAR WALL GUSSET PLATE AND HOLDOWN ASSEMBLY

E. DETAILS AT EXTERIOR WALLS

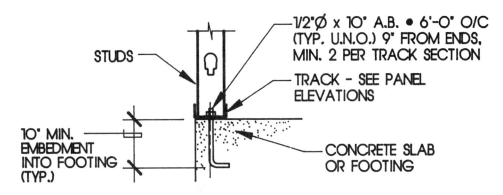

ANCHOR BOLT DETAIL

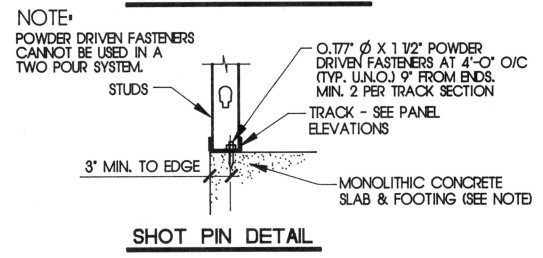

SHOT PIN DETAIL

TYPICAL TRACK ANCHORAGE DETAIL
AT EXTERIOR WALL ON SLAB

E.26

E. DETAILS AT EXTERIOR WALLS

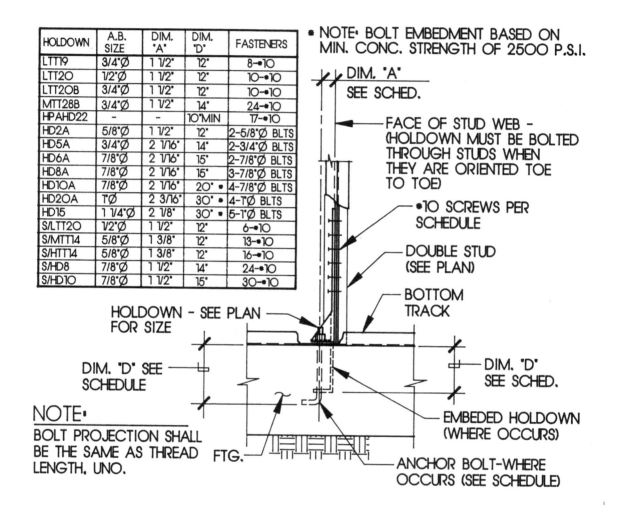

HOLDOWN	A.B. SIZE	DIM. "A"	DIM. "D"	FASTENERS
LTT19	3/4"Ø	1 1/2"	12"	8-#10
LTT20	1/2"Ø	1 1/2"	12"	10-#10
LTT20B	3/4"Ø	1 1/2"	12"	10-#10
MTT28B	3/4"Ø	1 1/2"	14"	24-#10
HPAHD22	-	-	10"MIN	17-#10
HD2A	5/8"Ø	1 1/2"	12"	2-5/8"Ø BLTS
HD5A	3/4"Ø	2 1/16"	14"	2-3/4"Ø BLTS
HD6A	7/8"Ø	2 1/16"	15"	2-7/8"Ø BLTS
HD8A	7/8"Ø	2 1/16"	15"	3-7/8"Ø BLTS
HD10A	7/8"Ø	2 1/16"	20" *	4-7/8"Ø BLTS
HD20A	1"Ø	2 3/16"	30" *	4-1"Ø BLTS
HD15	1 1/4"Ø	2 1/8"	30" *	5-1"Ø BLTS
S/LTT20	1/2"Ø	1 1/2"	12"	6-#10
S/MTT14	5/8"Ø	1 3/8"	12"	13-#10
S/HTT14	5/8"Ø	1 3/8"	12"	16-#10
S/HD8	7/8"Ø	1 1/2"	14"	24-#10
S/HD10	7/8"Ø	1 1/2"	15"	30-#10

* NOTE: BOLT EMBEDMENT BASED ON MIN. CONC. STRENGTH OF 2500 P.S.I.

NOTE:
BOLT PROJECTION SHALL BE THE SAME AS THREAD LENGTH, UNO.

E.27 TYPICAL HOLDOWN DETAIL AND SCHEDULE

E. DETAILS AT EXTERIOR WALLS

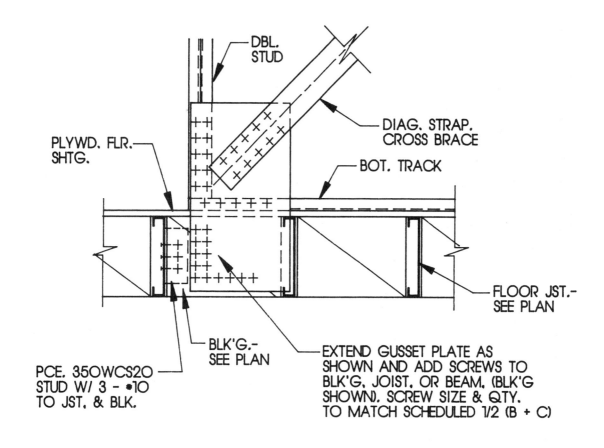

SECOND FLOOR SHEAR HOLDOWN TO FLOOR JOISTS

E.28

E. DETAILS AT EXTERIOR WALLS

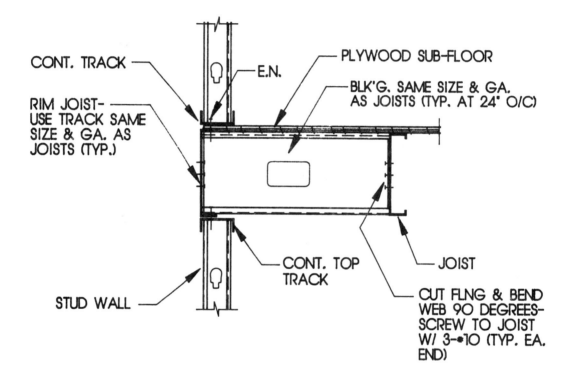

E.29 — EXTERIOR WALL SECTION WITH PARALLEL FLOOR JOISTS

E. DETAILS AT EXTERIOR WALLS

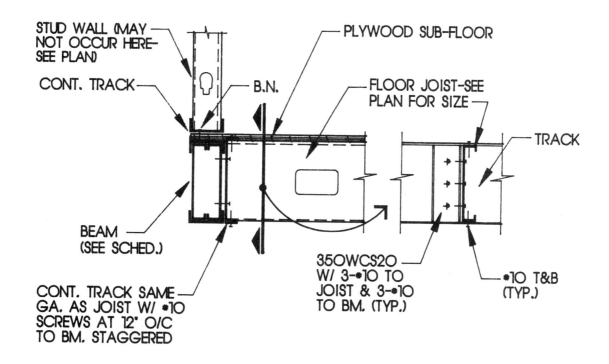

FLOOR JOIST TO FLUSH FRAMED BEAM CONNECTION

E.30

E. DETAILS AT EXTERIOR WALLS

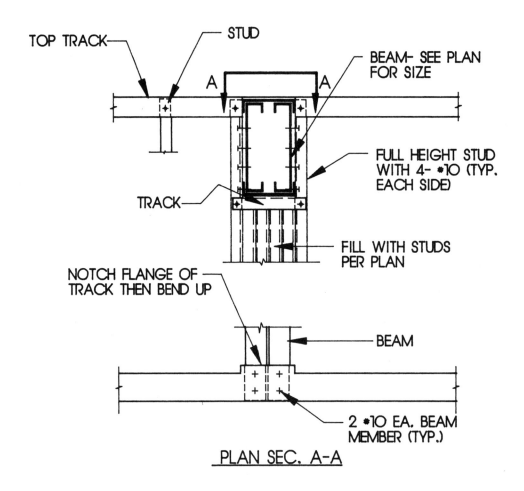

E.31   TYPICAL DROPPED BEAM TO WALL CONNECTION

E. DETAILS AT EXTERIOR WALLS

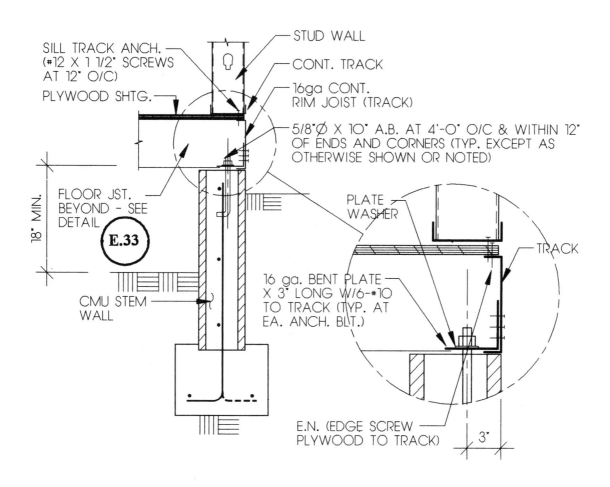

STEM WALL DETAIL — E.32

E. DETAILS AT EXTERIOR WALLS

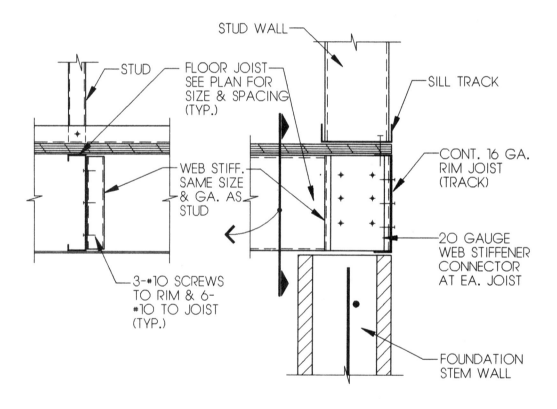

E.33 — JOIST CONNECTION AT STEM WALL

E. DETAILS AT EXTERIOR WALLS

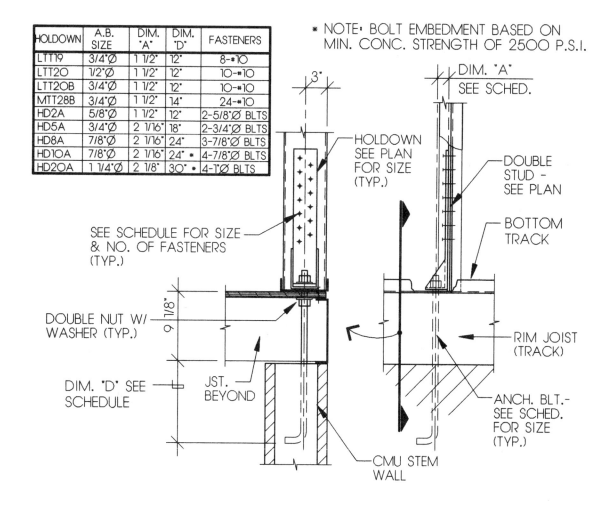

HOLDOWN	A.B. SIZE	DIM. "A"	DIM. "D"	FASTENERS
LTT19	3/4"Ø	1 1/2"	12"	8-#10
LTT20	1/2"Ø	1 1/2"	12"	10-#10
LTT20B	3/4"Ø	1 1/2"	12"	10-#10
MTT28B	3/4"Ø	1 1/2"	14"	24-#10
HD2A	5/8"Ø	1 1/2"	12"	2-5/8"Ø BLTS
HD5A	3/4"Ø	2 1/16"	18"	2-3/4"Ø BLTS
HD8A	7/8"Ø	2 1/16"	24"	3-7/8"Ø BLTS
HD10A	7/8"Ø	2 1/16"	24" *	4-7/8"Ø BLTS
HD20A	1 1/4"Ø	2 1/8"	30" *	4-1"Ø BLTS

* NOTE: BOLT EMBEDMENT BASED ON MIN. CONC. STRENGTH OF 2500 P.S.I.

HOLDOWN DETAIL AT STEM WALL E.34

E. DETAILS AT EXTERIOR WALLS

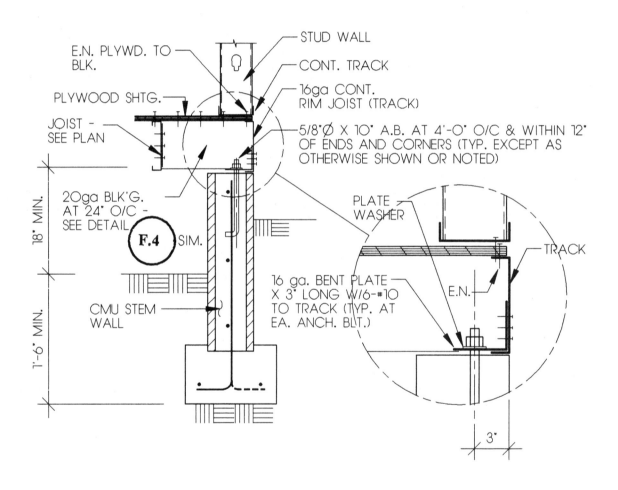

E.35 — PARALLEL FLOOR JOIST TO STEM WALL

F. MISC. BRIDGING, BLOCKING, REINFORCEMENT, ETC.

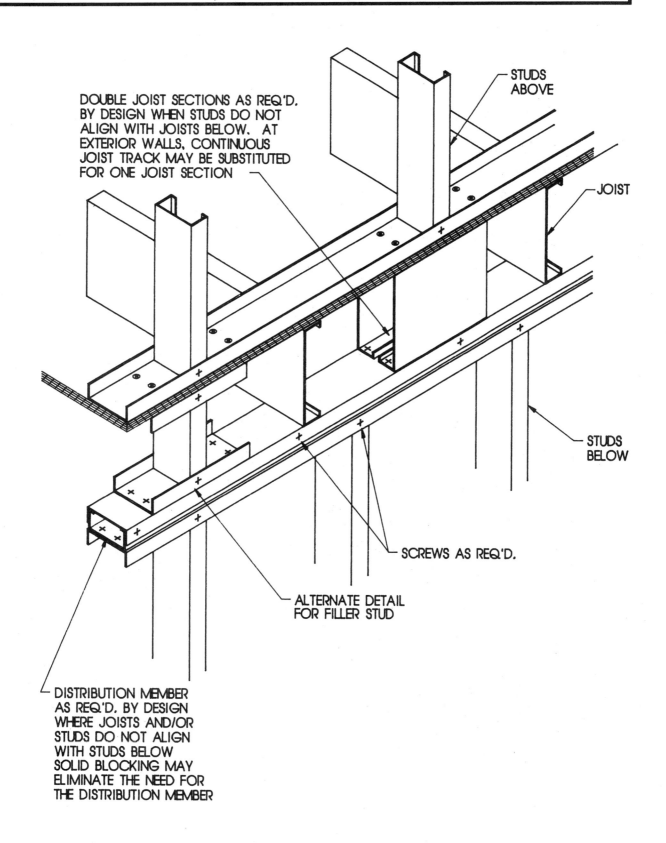

TOP TRACK LOAD DISTRIBUTION DETAILS F.1

F. MISC. BRIDGING, BLOCKING, REINFORCEMENT, ETC.

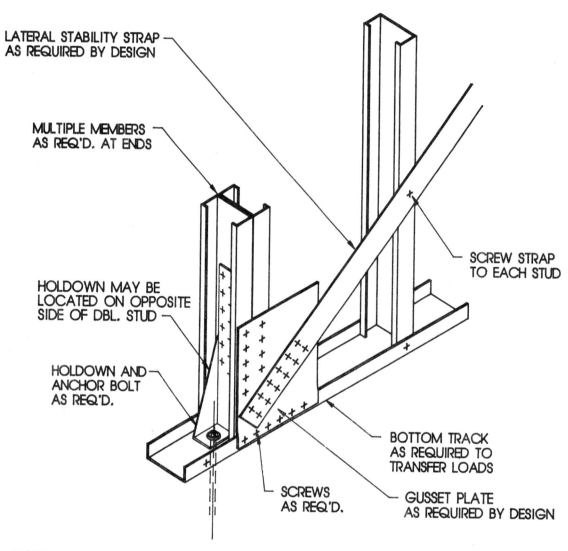

F.2   SHEAR WALL HOLDOWN AT BASE

F. MISC. BRIDGING, BLOCKING, REINFORCEMENT, ETC.

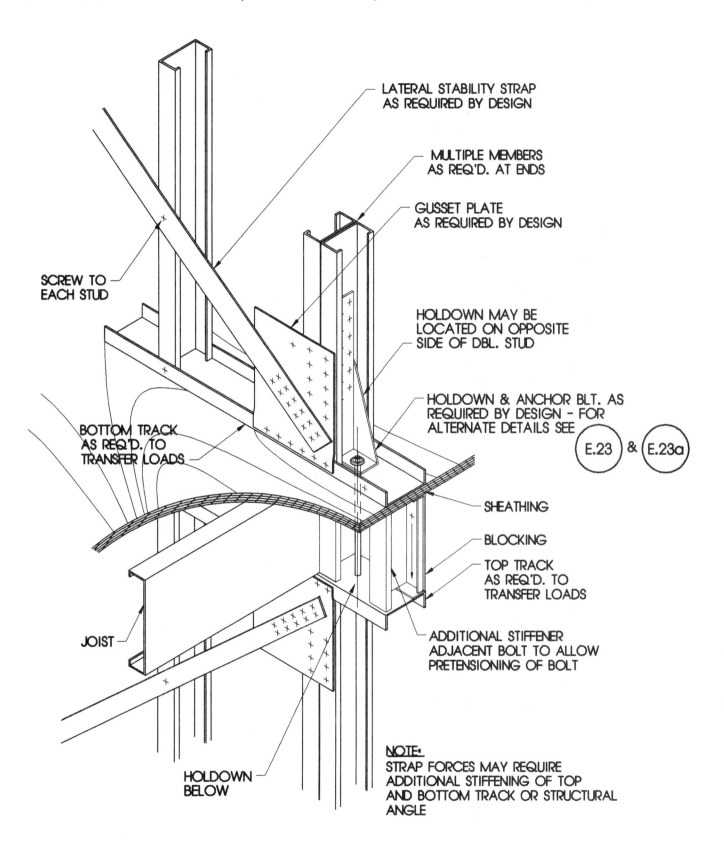

SHEAR WALL HOLDOWN AT SECOND FLOOR — F.3

F. MISC. BRIDGING, BLOCKING, REINFORCEMENT, ETC.

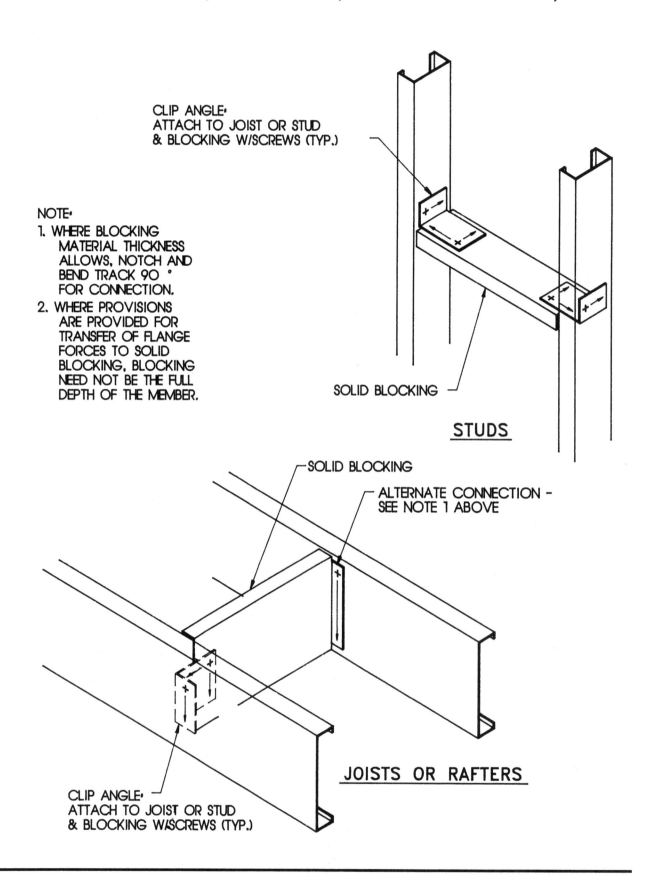

NOTE:
1. WHERE BLOCKING MATERIAL THICKNESS ALLOWS, NOTCH AND BEND TRACK 90° FOR CONNECTION.
2. WHERE PROVISIONS ARE PROVIDED FOR TRANSFER OF FLANGE FORCES TO SOLID BLOCKING, BLOCKING NEED NOT BE THE FULL DEPTH OF THE MEMBER.

F. MISC. BRIDGING, BLOCKING, REINFORCEMENT, ETC.

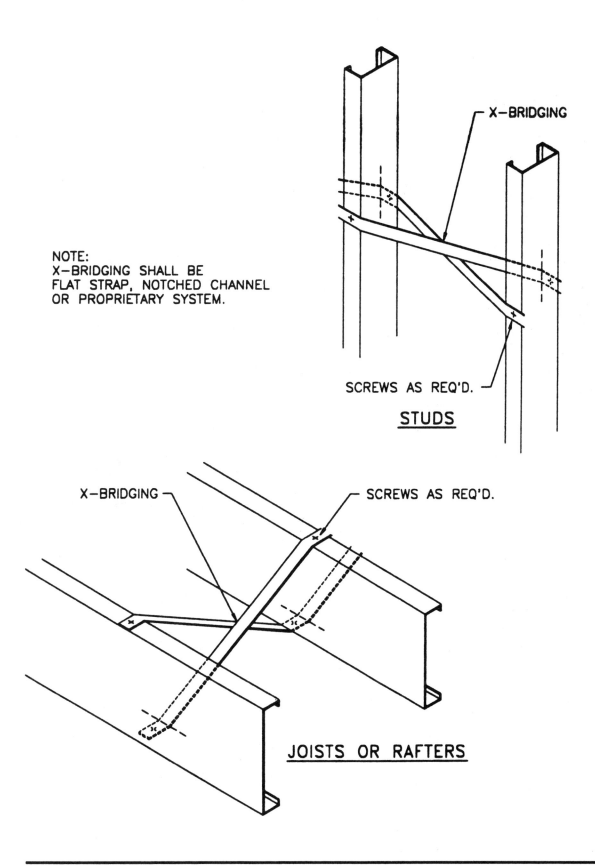

NOTE:
X-BRIDGING SHALL BE FLAT STRAP, NOTCHED CHANNEL OR PROPRIETARY SYSTEM.

CROSS BRIDGING   **F.5**

F. MISC. BRIDGING, BLOCKING, REINFORCEMENT, ETC.

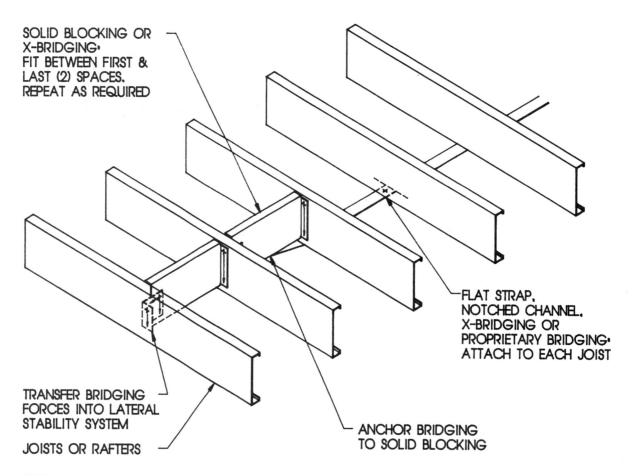

JOIST AND RAFTER BRIDGING

F. MISC. BRIDGING, BLOCKING, REINFORCEMENT, ETC.

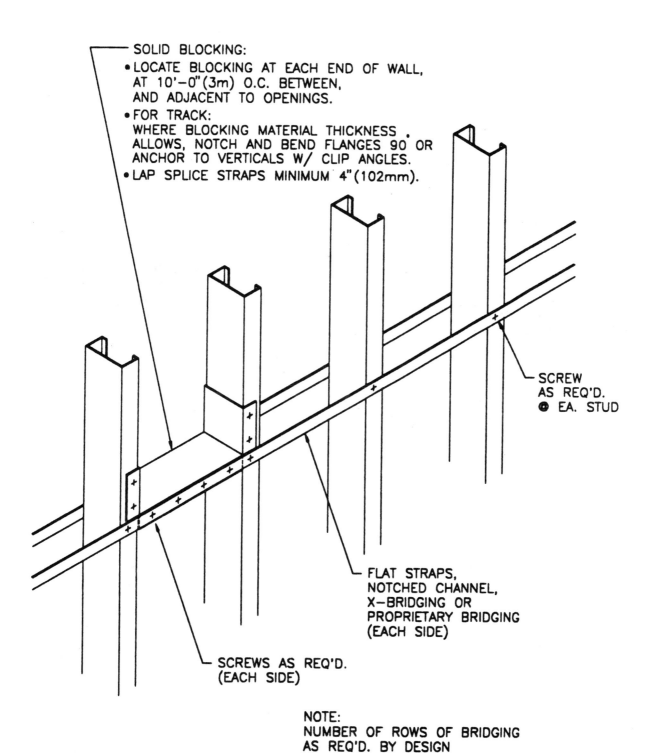

WALL BRIDGING   F.7

F. MISC. BRIDGING, BLOCKING, REINFORCEMENT, ETC.

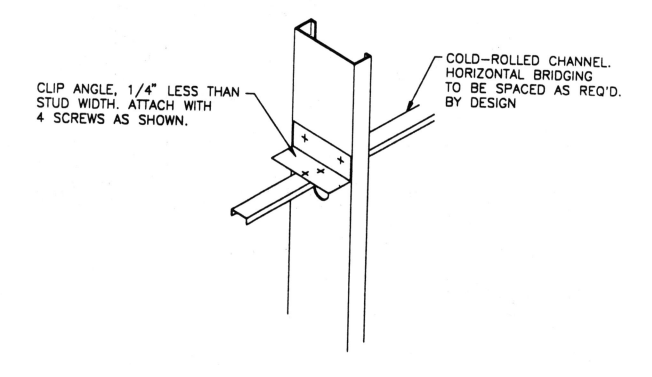

CLIP ANGLE, 1/4" LESS THAN STUD WIDTH. ATTACH WITH 4 SCREWS AS SHOWN.

COLD-ROLLED CHANNEL. HORIZONTAL BRIDGING TO BE SPACED AS REQ'D. BY DESIGN

F.8   WALL BRIDGING (ALTERNATE)

F. MISC. BRIDGING, BLOCKING, REINFORCEMENT, ETC.

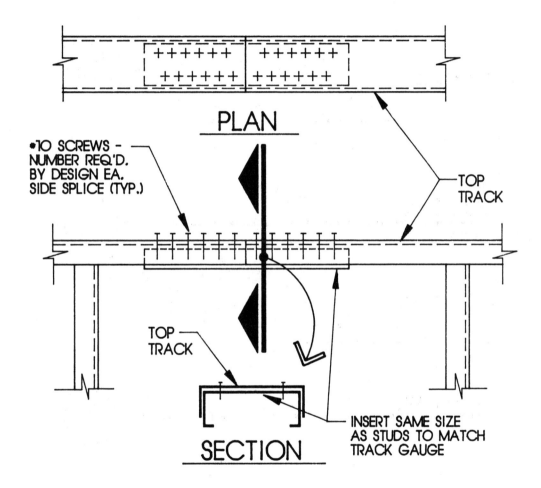

TOP TRACK SPLICE DETAIL — F.9a

F. MISC. BRIDGING, BLOCKING, REINFORCEMENT, ETC.

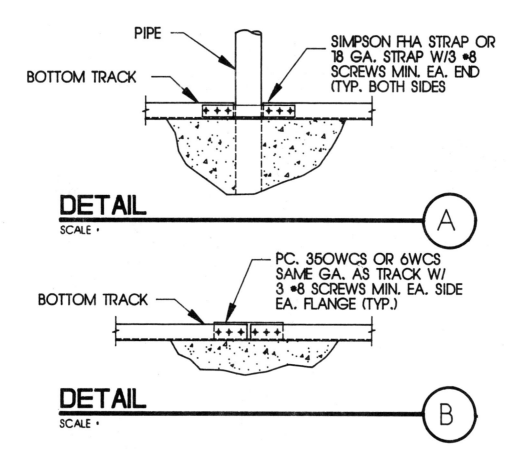

F.9b BOTTOM TRACK SPLICE DETAIL

F. MISC. BRIDGING, BLOCKING, REINFORCEMENT, ETC.

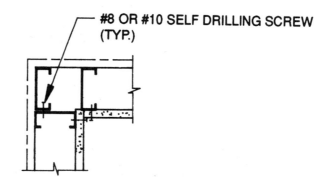

| F.10 | TYPICAL EXTERIOR CORNER FRAMING | F.10 |

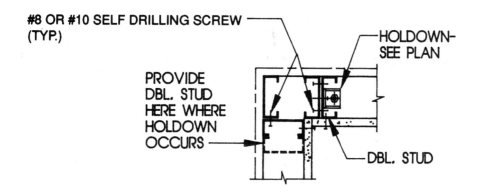

EXTERIOR CORNER FRAMING WITH HOLDOWN — F.10a

F. MISC. BRIDGING, BLOCKING, REINFORCEMENT, ETC.

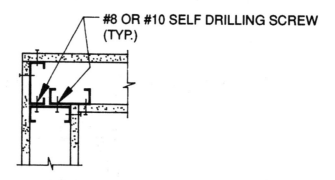

F.10b TYPICAL INTERIOR CORNER FRAMING **F.10b**

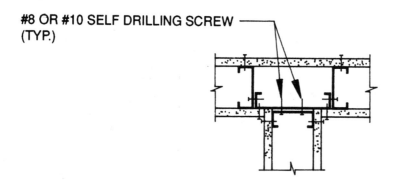

F.11 TYPICAL INTERIOR INTERSECTION **F.11**

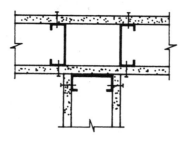

F.11a ALTERNATE INTERIOR INTERSECTION FRAMING

F. MISC. BRIDGING, BLOCKING, REINFORCEMENT, ETC.

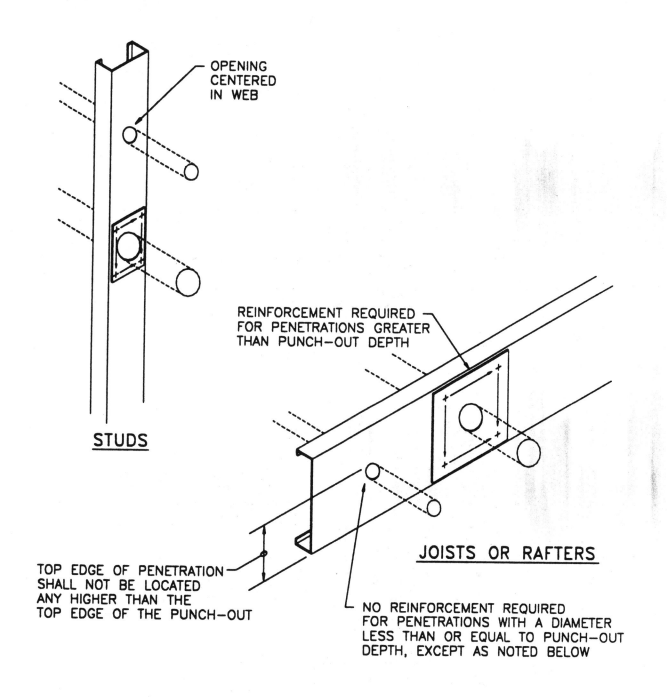

NOTES:
1. FLANGES SHALL NOT BE NOTCHED OR CUT.
2. CAPACITY VERIFICATION BY DESIGN IS REQ'D. FOR ANY OPENINGS LOCATED AT CONCENTRATED LOADS AND BEARING ENDS.

JOIST, STUD OR RAFTER WEB PENETRATIONS F.12

F. MISC. BRIDGING, BLOCKING, REINFORCEMENT, ETC.

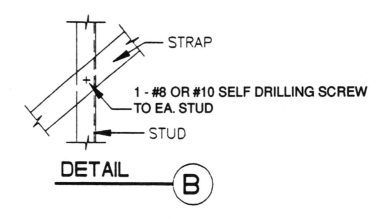

DETAIL B

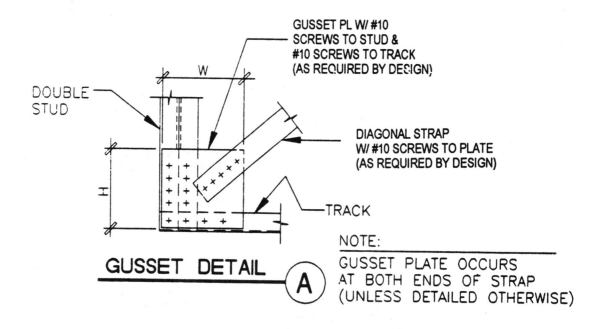

GUSSET DETAIL A

NOTE: GUSSET PLATE OCCURS AT BOTH ENDS OF STRAP (UNLESS DETAILED OTHERWISE)

F.13   SHEAR PANEL BRACING DETAILS

F. MISC. BRIDGING, BLOCKING, REINFORCEMENT, ETC.

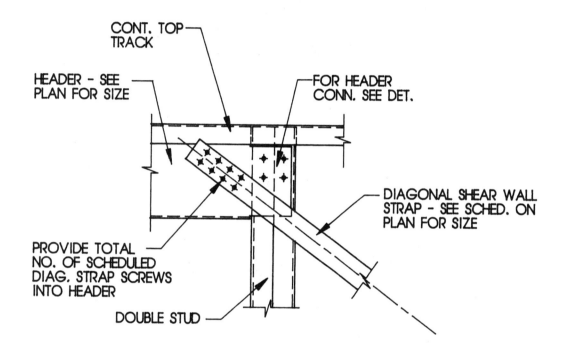

DIAGONAL STRAP ATTACHMENT TO HEADER   **F.13a**

F. MISC. BRIDGING, BLOCKING, REINFORCEMENT, ETC.

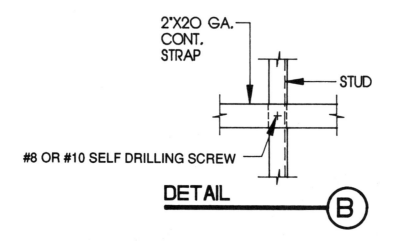

DETAIL B

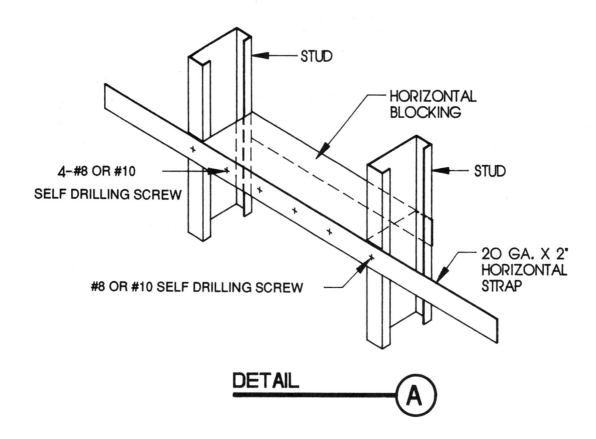

DETAIL A

F.14 WALL HORIZONTAL BLOCKING / BRIDGING DETAIL

F. MISC. BRIDGING, BLOCKING, REINFORCEMENT, ETC.

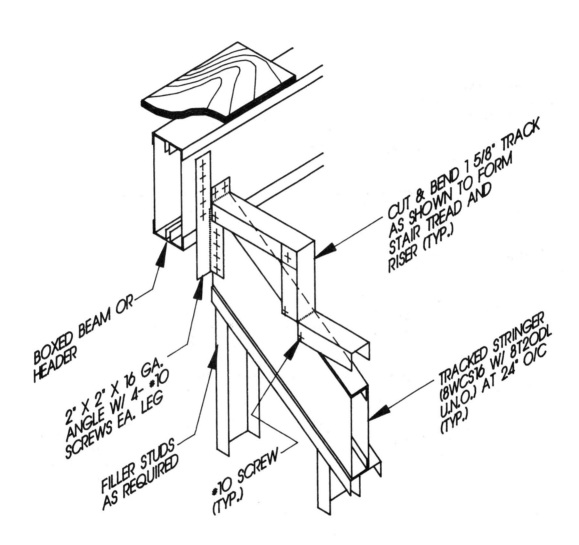

TYPICAL STAIR STRINGER CONNECTION F.15

F. MISC. BRIDGING, BLOCKING, REINFORCEMENT, ETC.

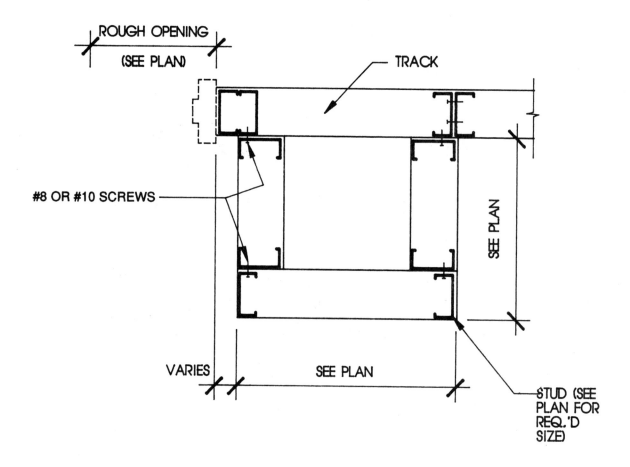

F.16 EXTERIOR POP-OUT DETAIL

F. MISC. BRIDGING, BLOCKING, REINFORCEMENT, ETC.

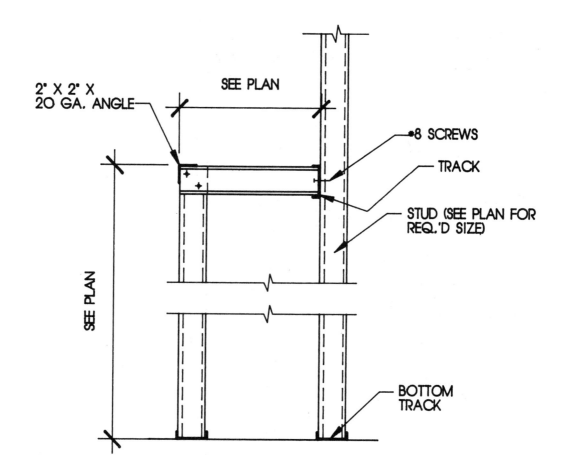

TYPICAL POT SHELF DETAIL   **F.17**

F. MISC. BRIDGING, BLOCKING, REINFORCEMENT, ETC.

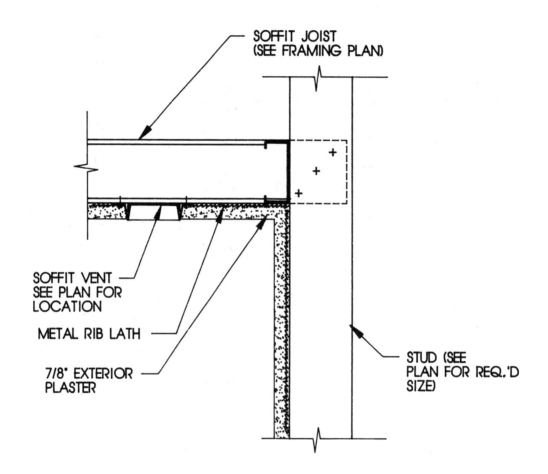

F.18   TYPICAL PLASTER SOFFIT DETAIL

F. MISC. BRIDGING, BLOCKING, REINFORCEMENT, ETC.

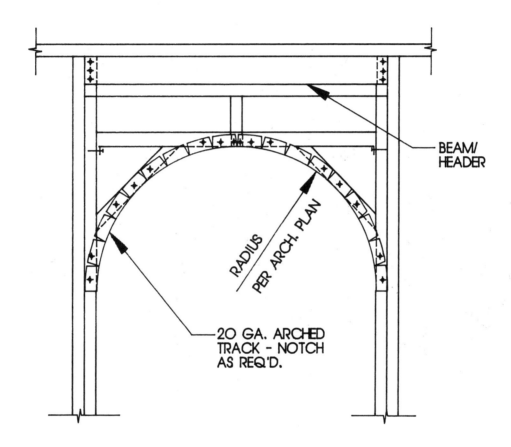

TYPICAL ARCH OPENING DETAIL   **F.19**

F. MISC. BRIDGING, BLOCKING, REINFORCEMENT, ETC.

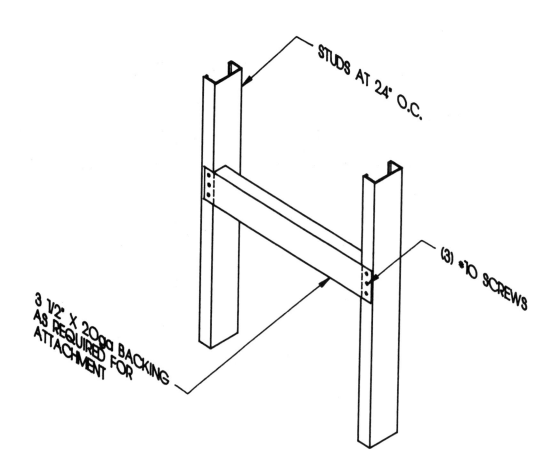

F.20   ATTACHMENT BACKING DETAIL

F. MISC. BRIDGING, BLOCKING, REINFORCEMENT, ETC.

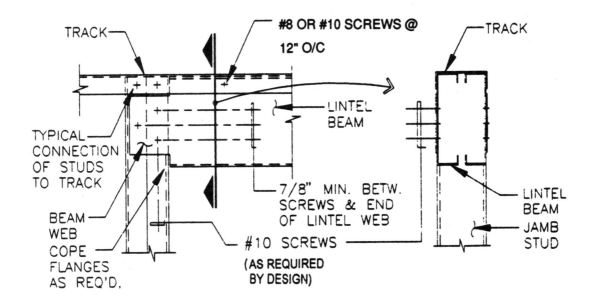

TYPICAL LINTEL BEAM CONNECTION

F.21

F. MISC. BRIDGING, BLOCKING, REINFORCEMENT, ETC.

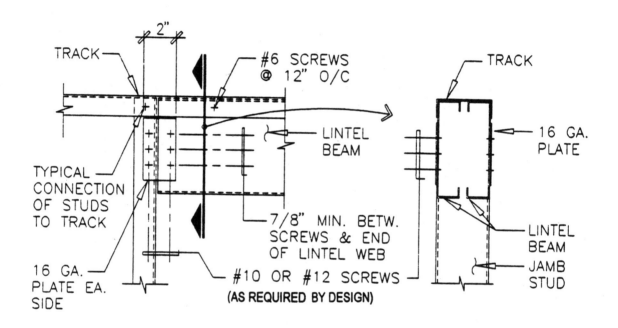

F.22   TYPICAL LINTEL BEAM CONNECTION (ALTERNATE)

F. MISC. BRIDGING, BLOCKING, REINFORCEMENT, ETC.

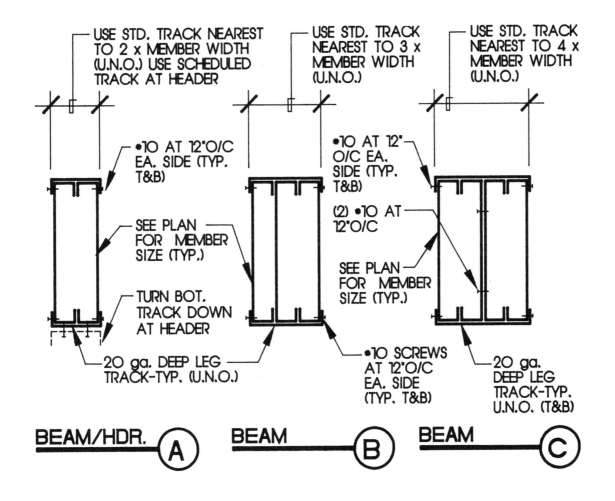

TYPICAL BOXED HEADER AND BEAM DETAILS F.23

F. MISC. BRIDGING, BLOCKING, REINFORCEMENT, ETC.

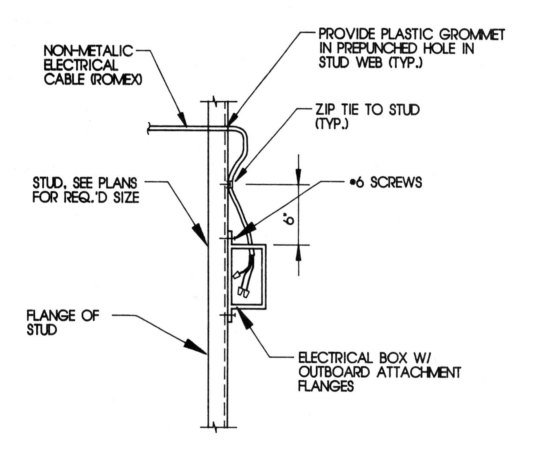

F.24 TYPICAL ELECTRICAL ATTACHMENT DETAIL

F. MISC. BRIDGING, BLOCKING, REINFORCEMENT, ETC.

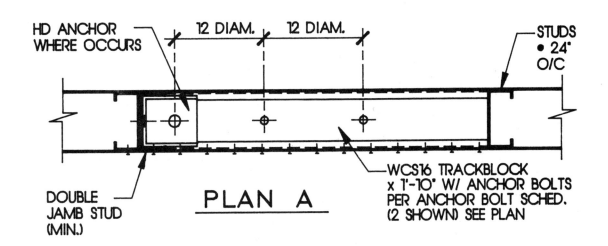

PLAN A

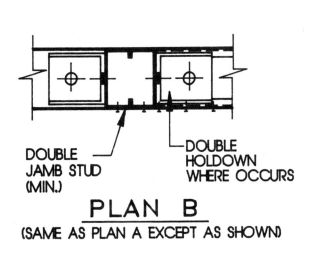

PLAN B
(SAME AS PLAN A EXCEPT AS SHOWN)

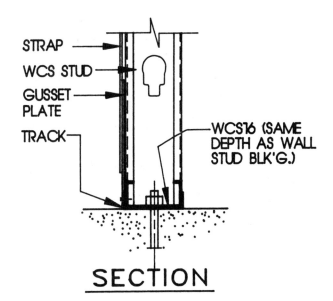

SECTION

SHEAR WALL TRACK REINFORCEMENT AND ANCHORAGE

F.25

F. MISC. BRIDGING, BLOCKING, REINFORCEMENT, ETC.

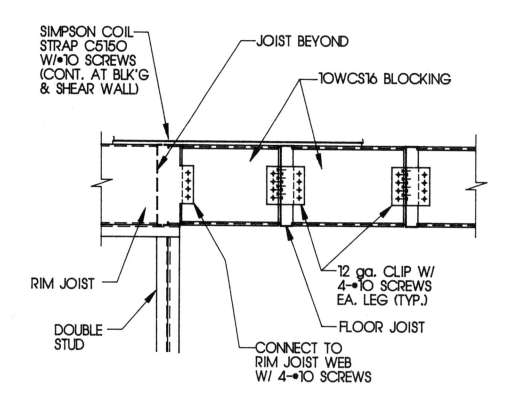

F.26 DRAG STRUT BLOCKING INTO FLOOR DIAPHRAGM

F. MISC. BRIDGING, BLOCKING, REINFORCEMENT, ETC.

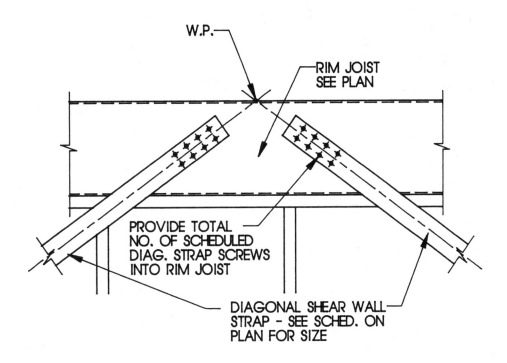

INVERTED CHEVRON TYPE STRAP CONNECTION TO RIM JOIST

F.27

F. MISC. BRIDGING, BLOCKING, REINFORCEMENT, ETC.

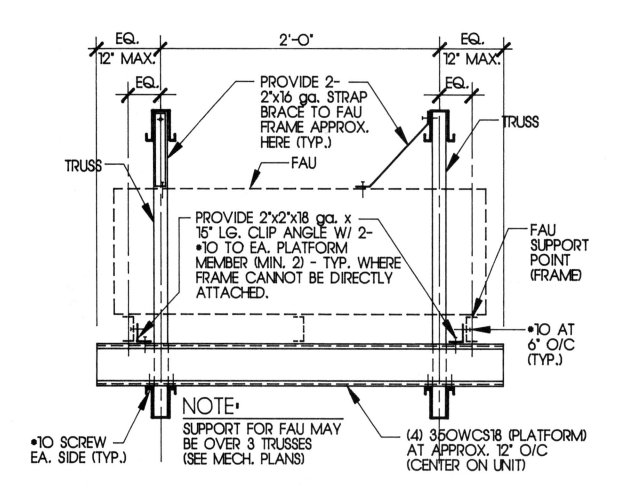

F.28 MECHANICAL UNIT SUPPORT BETWEEN TRUSSES IN ATTIC SPACE

G. FASTENING SCHEDULE RECOMMENDATIONS

Comparative Framing Designs (Wood vs. Light-Gauge Steel)

1. Design Conditions:
    a. Framed area, spans, supports and bracing.
    b. Rafters supporting drywall ceiling.
    c. Roof system sloped at 4:12

2. Design Criteria:
    a. Live Load 20 lb/ft^2
    b. Dead Load: 18 lb./ft^2
    c. Limiting Deflection (live load only):
       1/240 clear-span.

Roof Framing Comparison

Wood	Framing Item	Light-Gauge Steel
#2 kiln-dried Spruce-Pine-Fir 2x10 @ 16" O.C., notched to bear on ridge beam, 2x12 ridge board over beam, 16d common wire nails.	Rafter Size / Spacing / Span Condition	6", 18 ga. rafter at 24" O.C. full-length, lapped at ridge over beam support, #10 self drilling screws.
Rafter notched to bear on wall, toe-nailed with 8d common wire nails or attached by manufactured hurricane ties.	Exterior Wall Attachment	Connection clip with ¾" # 10 self drilling screws.
2x10 sub-fascia, 16d common wire nails; 1x12 fascia, 8d common wire nails.	Rafter End Closure	6", 18 ga. Rim Track with ¾" #10 Low Profile Head Screws to Web Stiffener and from Web Stiffener to rafter, 1x8 Wood Fascia to Rim Track with 1 ⅛" screw-shank nails.
8d common wire nails, toe-nailed.	Rafter to Ridge Beam Attachment	Connection clips, ⅝" #10 Low profile self drilling screws.
½" CDX plywood, 8d common wire nails	Roof Sheathing, and Attachment to Rafters	½" CDX plywood, 1" #8 or #10 counter sinking head self drilling screws.

The above data has been provided for comparative purposes only and must *not* be used for design. Consult manufacturers' data for design values and specifications.

ROOF DESIGN (COMPARATIVE FRAMING DESIGNS)

G. FASTENING SCHEDULE RECOMMENDATIONS

Recommended Fasteners in Floor Framing

Materials	Fastener	Frequency or Quantity
Joist to Wood Sill	1" #10 Self Drilling Low Profile	1 @ each Joist to Wood Sill
Joist to Girder	½" #10 Self Drilling Low Profile	1 @ Joist to Girder
Joist to Connection Clip	½" #10 Self Drilling Low Profile	3 to 4 @ each clip
Bridging to Joist	½" #10 Self Drilling Low Profile	1 @ each joist
Joist to 2x Wood End Stiffener	1 ¼" #10 Self Drilling Low Profile	1 @ each Joist Web to Stiffener
End Stiffener to Joist	¾" #10 Self Drilling Low Profile	3 to 4 @ each Stiffener to Joist
End Stiffener to Wood Rim Joist	8d Common Wire Nail	2 @ each End Stiffener to Rim Joist
Steel Rim Joist to End Stiffener	¾" #10 Self Drilling Low Profile	3 @ each Joist
Steel Rim Track to End Stiffener	¾" #10 Self Drilling Low Profile	3 @ each Joist
Joist Hanger to Joist	⅞" #10 Self Drilling Low Profile	3 @ each Joist
Joist to Overlapping Joist	¾" #10 Self Drilling Low Profile	3 @ each Support
Wood Sole Plate (Wall) to Rim Joist & Track	2 ½" #10 or #12 Self Drilling Low Profile	1 @ 24" O.C. & Max. 12" from each end of Track
Plate Track (Bottom) to Joist & Track	1 ¹⁵⁄₁₆" #10 or #12 Self Drilling Low Profile	1 @ 24" O.C. & Max. 12" from each end of Track

Notes:
(1) Low Profile Head is used in lieu of Pan or Hexwasher heads where least projection of fastener is desired.
(2) S-7 point will substitute S-12 when attaching .07" members together.
(3) #2 point self drilling screw will be substituted by #3 point self drilling screw when steel thickness varies between .09" to .250". Consult MFG. recommended thickness for drill capacity.

The above data has been provided for comparative purposes only and must *not* be used for design. Consult manufacturers' data for design values and specifications.

FLOOR FRAMING

G. FASTENING SCHEDULE RECOMMENDATIONS

Recommended Fasteners in Wall Framing (Load Bearing)

Materials	Fastener	Frequency or Quantity
Stud to Plate Track (Bottom)	⅝" - ¾" #8 or #10 Self Drilling Low Profile Head	1 @ each Flange
Stud to Plate Track (Top)	⅝" - ¾" #8 or #10 Self Drilling Low Profile Head	1 @ each Flange
Diagonal Bracing to Stud	½" - ⅝" #8 or #10 Self Drilling Low Profile Head	1 @ each Stud
Lateral Bracing to Stud	½" - ⅝" #8 or #10 Self Drilling Low Profile Head	1 @ each Stud Per Strap or 3@ each connection clip with Cold Rolled Channel
Gusset to Stud	½" - ⅝" #8 or #10 Self Drilling Low Profile Head	Quantity and Spacing as per Loading
Stud to Stud (Nested)	½" - ⅝" #8 or #10 Self Drilling Low Profile Head	1 @ 24" O.C. through Flange
Stud to Stud (Back to Back)	½" - ⅝" #8 or #10 Self Drilling Low Profile Head	1 @ 24" O.C. through Web
Stud to Stud (@ Wall Intersection)	½" - ⅝" #8 or #10 Self Drilling Low Profile Head	1 @ 24" O.C. or 1 @ each Blocking
Lintel to Stud	½" - ⅝" #8 or #10 Self Drilling Low Profile Head	Requirement Varies with Different Loading

Notes:
(1) Low Profile Head is used in lieu of Pan or Hexwasher heads where least projection of fastener is desired.
(2) S-7 point will substitute S-12 when attaching .07" members together.
(3) #2 point self drilling screw will be substituted by #3 point self drilling screw when steel thickness varies between .09" to .250". Consult MFG. recommended thickness for drill capacity.

The above data has been provided for comparative purposes only and must *not* be used for design. Consult manufacturers' data for design values and specifications.

WALL FRAMING (LOAD BEARING)

G. FASTENING SCHEDULE RECOMMENDATIONS

Recommended Fasteners in Wall Framing (Non-Load-Bearing / Drywall)

Materials	Fastener	Frequency or Quantity
Stud to Plate Track (Bottom)	½" - ⅝" #8 or #10 Self Drilling Low Profile	1 @ each Flange
Stud to Plate Track (Top)	½" #8 Self Drilling Low Profile	1 @ each Flange
Lateral Bracing to Stud	½" #8 Self Drilling Low Profile	2 @ Flange
Stud to Stud (Nested)	½" #10 Self Drilling Low Profile	1 @ 24" O.C.
Stud to Stud (Back to Back)	½" #10 Self Drilling Low Profile	1 @ 24" O.C.
Stud to Stud (@ Wall Intersection)	½" #10 Self Drilling Low Profile	1 @ 24" O.C. or 1 @ each Blocking

Notes:
(1) Low Profile Head is used in lieu of Pan or Hexwasher heads where least projection of fastener is desired.
(2) S-7 point will substitute S-12 when attaching .07" members together.
(3) #2 point self drilling screw will be substituted by #3 point self drilling screw when steel thickness varies between .09" to .250". Consult MFG. recommended thickness for drill capacity.

The above data has been provided for comparative purposes only and must *not* be used for design. Consult manufacturers' data for design values and specifications.

G. FASTENING SCHEDULE RECOMMENDATIONS

Recommended Fasteners in Roof Framing

Materials	Fastener	Frequency or Quantity
Ceiling Joist to Wood Top Plate	1"- 1 ⅛" #10 or #12 Self Piercing or Type 17 point	1 @ each Joist
Ceiling Joist to Top Plate Track	⅝" - ¾" #10 Self Drilling Pan Head	1 @ each Joist
Connection Clip to Top Plate Track	1"- 1 ⅛" #10 or #12 Self Piercing or Type 17 point	4 @ each Clip to Top Plate
Connection Clip to Top Plate Track	⅝" - ¾" #10 Self Drilling Pan Head	4 @ each Clip to Plate Track
Connection Clip to Ceiling Joist	⅝" - ¾" #10 Self Drilling Pan Head	Min. 3 @ each Clip to Ceiling Joist and as per Loading
Connection Clip to Rafter	⅝" - ¾" #10 Self Drilling Pan Head	Min. 3 @ each Clip to Rafter and as per Loading
Ceiling Joist to Parallel Rafter	⅝" - ¾" #10 Self Drilling Pan Head	No. varies as per loading
Ceiling Joist to Truss Web	⅝" - ¾" #10 Self Drilling Pan Head	Min. 2 @ Flange and as per Loading Joist
Ceiling Joist, Overlapped at Support	⅝" - ¾" #10 Self Drilling Pan Head	Min. 2 @ Web
Connection Clip to Ridge Board	⅝" - ¾" #10 Self Drilling Pan Head	4 - 6 @ each Clip to Ridge
Rafters Overlapped at Ridge	⅝" - ¾" #10 Self Drilling Pan Head	Min. 6 @ Overlapped Web Section and as per Loading
Built up Beam (Ridge Board)	⅝" - ¾" #10 Self Drilling Pan Head	1 @ each Flange @ 12" O.C.
Stiffback Bracing to Joist	⅝" - ¾" #10 Self Drilling Pan Head	Min. 2 @ each Joist
Sub-Fascia Track to Rafter	⅝" - ¾" #10 Self Drilling Low Profile Pan Head	1 @ each Connection Clip and Max Top Plate
Wood Fascia to Sub-Fascia Track	1 ⅝"#6 Trim Head	2 @ 24" O.C. and @ maximum of 12" from Each End of Board or Corner

Notes:
(1) Low Profile Head is used in lieu of Pan or Hexwasher heads where least projection of fastener is desired.
(2) S-7 point will substitute S-12 when attaching .07" members together.
(3) #2 point self drilling screw will be substituted by #3 point self drilling screw when steel thickness varies between .09" to .250". Consult MFG. recommended thickness for drill capacity.

The above data has been provided for comparative purposes only and must *not* be used for design. Consult manufacturers' data for design values and specifications.

ROOF FRAMING

G. FASTENING SCHEDULE RECOMMENDATIONS

Recommended Fasteners in Application of Single Ply Gypsum Wall Board.

Thickness of Gypsum Wallboard	Plane of Framing Surface	Long Dimension Gypsum Board in Relation to Direction of Framing Member	Max. Spacing of Framing Member	Screw Type & Size			Maximum Spacing of Fasteners Screw Shank Nail				
					Edge	Field	Type & Size	without Adhesive Edge	Field	with Adhesive Edge	Field
⅜" - ½"	Horizontal	Ether Direction	16	1" D/W Self Drilling or Sharp Point	6	12	1 ¼" Screw Shank Nail	6	8	16	16
	Horizontal	Perpendicular	24		6	12		6	8	16	16
	Vertical	Ether Direction	24		6	12		8	8	16	16
⅝" - ¾"	Horizontal	Ether Direction	16	1" - 1 ¼" D/W Self Drilling or Sharp Point	6	12	1 ⅝" Screw Shank Nail	6	8	16	16
	Horizontal	Perpendicular	24		6	12		6	8	16	16
	Vertical	Ether Direction	24		6	12		8	8	16	16

Notes:
(1) Gyp Board to 20 ga. & 25 ga. should use a sharp point drywall screw.
(2) Gyp Board to two (2) thicknesses of 20 ga. or up to 14 ga. should use a self-drilling drywall screw.

The above data has been provided for comparative purposes only and must *not* be used for design. Consult manufacturers' data for design values and specifications.

G. FASTENING SCHEDULE RECOMMENDATIONS

Recommended Fasteners in Application of Single Ply Plywood for Combination Subfloor - Underlayment to Framing.

Thickness of Plywood Wallboard	Plane of Framing Surface	Long Dimension Plywood Board in Relation to Direction of Framing Member	Max. Spacing of Framing Member	Screw Type & Size	Maximum Spacing of Fasteners - Screw Edge	Field	Screw Shank Nail Type & Size	without Adhesive Edge	Field	with Adhesive Edge	Field
½" & Less	Horizontal	Ether Direction	16	1 ¼" - 1 ⁷⁄₁₆" #10 Self Drilling waferhead	6	12	1 1/8" Screw Shank Nail	6	12	6	12
	Horizontal	Perpendicular	24		6	12		6	12	6	12
	Vertical	Ether Direction	24		6	12		6	12	6	12
⅝" & ¾"	Horizontal	Ether Direction	16	2" #8 Pilot Self Drilling or 1 ⁷⁄₁₆" #10 Winged Self Drilling	6	12	1 5/8" Screw Shank Nail	6	12	6	12
	Horizontal	Perpendicular	24		6	12		6	12	6	12
	Vertical	Ether Direction	24		6	12		6	12	6	12
⅞" - 1"	Horizontal	Ether Direction	16	1 ⅝" #10 Flat Winged Self Drilling or 2 ½" #12 Flat Pilot Self Drilling	6	12	1 7/8" Screw Shank Nail	6	12	6	12
	Horizontal	Perpendicular	24		6	12		6	12	6	12
	Vertical	Ether Direction	24		6	12		6	12	6	12
1 ⅛"	Horizontal	Ether Direction	16	2 ½" #10 or #12 Pilot Self Drilling	6	12	2 1/4" Screw Shank Nail	6	12	6	12
	Horizontal	Perpendicular	24		6	12		6	12	6	12
	Vertical	Ether Direction	24		6	12		6	12	6	12
1 ¼" - 1 ⅜"	Horizontal	Ether Direction	16	2 ¾" #14 Flat Winged Self Drilling	6	12	2 1/4" Screw Shank Nail	6	12	6	12
	Horizontal	Perpendicular	24		6	12		6	12	6	12
	Vertical	Ether Direction	24		6	12		6	12	6	12

Notes:
(1) If nails are used in lieu of screws, use of adhesive is strongly recommended.
(2) Winged screws are not recommended in 20 ga. or less.

The above data has been provided for comparative purposes only and must *not* be used for design. Consult manufacturers' data for design values and specifications.

PLYWOOD OR O.S.B. SHEATHING (SINGLE PLY)

G. FASTENING SCHEDULE RECOMMENDATIONS

Recommended Fasteners in Application of Single Ply Plywood for Combination Subfloor - Underlayment to Framing.

Thickness of Gypsum Plywood Subfloor	Plane of Framing Surface	Long Dimension Plywood Subfloor in Relation to Direction of Framing Member	Max. Spacing of Framing Member	Screw Type & Size	Maximum Spacing of Fasteners – Screw Edge	Field	Screw Shank Nail Type & Size	without Adhesive Edge	Field	with Adhesive Edge	Field
5/8" - 3/4"	Horizontal	Ether Direction	16	2" #8 Pilot Self Drilling or 1 7/16" #10 Winged Self Drilling	6	12	1 5/8" Screw Shank Nail	6	12	6	12
	Horizontal	Perpendicular	24		6	12		6	12	6	12
	Vertical	Ether Direction	24		6	12		6	12	6	12
7/8" - 1"	Horizontal	Ether Direction	16	1 7/8" #10 Flat Winged Self Drilling or 2 1/2" #12 Flat Pilot Self Drilling	6	12	1 7/8" Screw Shank Nail	6	12	6	12
	Horizontal	Perpendicular	24		6	12		6	12	6	12
	Vertical	Ether Direction	24		6	12		6	12	6	12
1 1/8" - 1 1/4"	Horizontal	Ether Direction	16	2 1/2" #10 or #12 Pilot Self Drilling	6	12	2 1/4" Screw Shank Nail	6	12	6	12
	Horizontal	Perpendicular	24		6	12		6	12	6	12
	Vertical	Ether Direction	24		6	12		6	12	6	12
1 1/4" - 1 3/8"	Horizontal	Ether Direction	16	2 3/4" #14 Flat Winged Self Drilling	6	12	2 1/4" Screw Shank Nail	6	12	6	12
	Horizontal	Perpendicular	24		6	12		6	12	6	12
	Vertical	Ether Direction	24		6	12		6	12	6	12

Notes:
(1) If nails are used in lieu of screws, use of adhesive is strongly recommended.
(2) Winged screws are not recommended in 20 ga. or less.
(3) Screw length must be increased to a minimum of underlayment thickness.

The above data has been provided for comparative purposes only and must *not* be used for design. Consult manufacturers' data for design values and specifications.

PLYWOOD OR O.S.B. FOR COMBINATION SUBFLOOR

H. FASTENERS

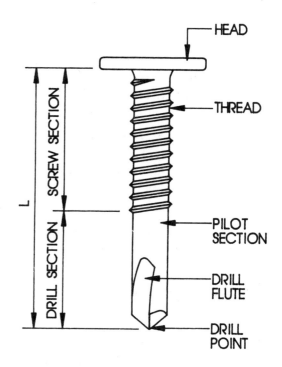

Applications

Floor Systems	A. Metal Lath to Joist B. Joist to Wood Web Stiffener C. Joist to Wood End Stiffener D. Joist to Wood Sill E. End Stiffener to Rim	F. Connection Clip to Joist G. Joist to Web Stiffener H. Bridging to Joist I. Solid Blocking to Joist J. Built-up Beam Assembly K. Joist Hanger to Joist L. Joist to Overlapping Joist
Wall Systems	A. Metal Lath to Stud B. Stud to Wood Web Stiffener C. Stud to Plate Track D. Ladder-Back to Stud E. Stud to Stud	F. Windbrace to Stud G. Lateral Bracing to Stud H. Gusset to Stud & Header I. Window Sill to Stud J. Lintel Assembly
Roof Systems	A. Metal Lath to Rafter B. Rafter to Wood Web Stiffener C. Ceiling Joist to Wood End Stiffener D. Subfascia to Rafter E. Rafter to Rafter F. Collar Tie to Rafter	G. Bridging to Rafter H. Rafter to Ceiling Joist I. Gusset to Rafter J. King Post to Rafter K. Truss Web to Rafter L. Bracing to Rafter M. Rafter to Web Stiffener

CHECK WITH MANUFACTURER FOR STANDARD AVAILABLE SIZES.

H.1   PANCAKE (LOW-PROFILE) HEAD SCREW   H.1

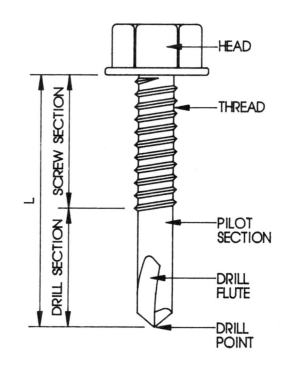

Applications

Floor Systems	A. End Stiffener to Rim B. Connection Clip to Joist C. Joist to Web Stiffener D. Bridging to Joist	E. Solid Blocking to Joist F. Built-up Beam Assembly G. Joist Hanger to Joist H. Joist to Overlapping Joist
Wall Systems	A. Stud to Plate Track B. Ladder-Back to Stud C. Stud to Stud D. Windbrace to Stud	E. Lateral Bracing to Stud F. Gusset to Stud & Header G. Window Sill to Stud H. Lintel Assembly
Roof Systems	A. Subfascia to Rafter B. Rafter to Rafter C. Collar Tie to Rafter D. Bridging to Rafter E. Rafter to Ceiling Joist	F. Gusset to Rafter G. King Post to Rafter H. Truss Web to Rafter I. Bracing to Rafter J. Rafter to Web Stiffener

CHECK WITH MANUFACTURER FOR STANDARD AVAILABLE SIZES.

H.2   HEX WASHER HEAD SCREW   H.2

H. FASTENERS

	Applications	
Floor Systems	A. End Stiffener to Rim B. Connection Clip to Joist C. Joist to Web Stiffener D. Bridging to Joist E. Solid Blocking to Joist	F. Built-up Beam Assembly G. Joist Hanger to Joist H. Joist to Overlapping Joist I. Trim Moulding to Joist
Wall Systems	A. Stud to Plate Track B. Ladder-Back to Stud C. Stud to Stud D. Windbrace to Stud E. Lateral Bracing to Stud	F. Gusset to Stud & Header G. Window Sill to Stud H. Lintel Assembly I. Wood Cabinet to Stud J. Trim Moulding to Stud
Roof Systems	A. Subfascia to Rafter B. Rafter to Rafter C. Collar Tie to Rafter D. Bridging to Rafter E. Rafter to Ceiling Joist F. Gusset to Rafter	G. King Post to Rafter H. Truss Web to Rafter I. Bracing to Rafter J. Rafter to Web Stiffener K. Trim Moulding to Rafter

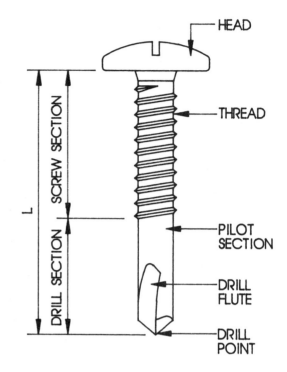

CHECK WITH MANUFACTURER FOR STANDARD AVAILABLE SIZES.

H.3 PAN HEAD SCREW

Applications
Wall Systems: A. Stud to Plate Track B. Ladder-Back to Stud C. Joist Hangers to Steel Beam D. Stud to Stud E. Window Sill to Stud F. Stud to Web Stiffener G. Track to Track

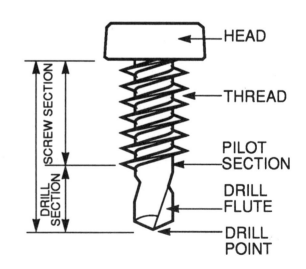

CHECK WITH MANUFACTURER FOR STANDARD AVAILABLE SIZES.
* AVAILABLE IN SHARP POINT AND SELF DRILLING (AS SHOWN)

H.4 PAN FRAMING SCREW

H. FASTENERS

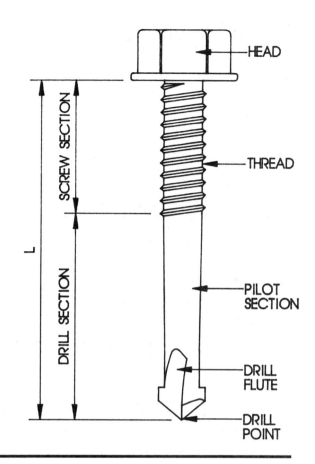

Applications

Floor Systems
- A. Floor Joists to Steel Beam
- B. Angle Clips or Angles to Steel Beam
- C. Joist Hangers to Steel Beam
- D. Rim Track to Steel Beam

CHECK WITH MANUFACTURER FOR STANDARD AVAILABLE SIZES.
* OTHER DRILL POINT SIZES AVAILABLE

H.5   HEX HEAD SELF-DRILLING #5 SCREW   **H.5**

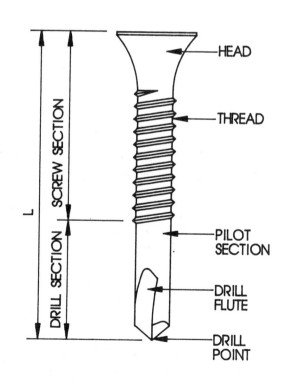

Applications

Floor Systems
- A. Joist to Wood Sill
- B. Sheathing to Joist
- C. Subflooring to Joist

Wall Systems
- A. Drywell to Stud
- B. Sheathing to Stud
- C. Furring to Lath
- D. Rigid Insulation to Stud

Roof Systems
- A. Drywall to Rafter
- B. Sheathing to Rafter
- C. Drywell to Ceiling Joist
- D. Rigid Insulation to Rafter

CHECK WITH MANUFACTURER FOR STANDARD AVAILABLE SIZES.
* AVAILABLE IN SHARP POINT AND SELF DRILLING (AS SHOWN)

BUGLE HEAD SCREW   **H.6**

H. FASTENERS

Applications	
Floor Systems	A. Trim Moulding to Joist
Wall Systems	A. Wood Trim to Stud
Roof Systems	A. Trim Attachment

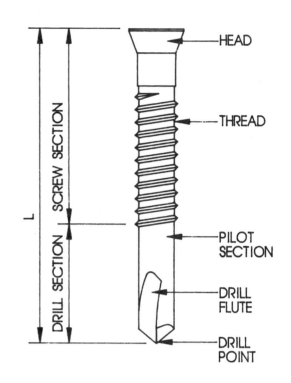

CHECK WITH MANUFACTURER FOR STANDARD AVAILABLE SIZES.
* AVAILABLE IN SHARP POINT AND SELF DRILLING (AS SHOWN)

H.7   TRIM HEAD SCREW   **H.7**

Applications	
Floor Systems	A. Wood Sheathing to Joists
Wall System	A. Wood Sheathing to Studs and Track
Roof Systems	A. Wood Sheathing to Trusses B. Wood Sheathing to Joists / Rafters

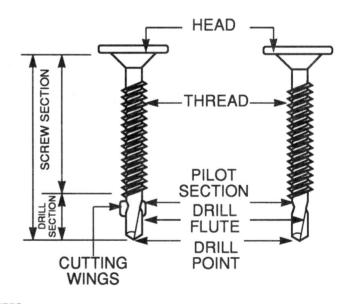

CHECK WITH MANUFACTURER FOR STANDARD AVAILABLE SIZES.
* ⅛" wood or less to 18 gauge through 12 gauge
** ¾" wood or less to 18 gauge through 12 gauge

H.8   WAFER HEAD OR WAFER WINGED SCREW   **H.8**

H. FASTENERS

Applications	
Floor Systems	A. Wood Sheathing to Joists
Wall System	A. Wood Sheathing to Studs and Track
Roof Systems	A. Wood Sheathing to Trusses B. Wood Sheathing to Joists / Rafters

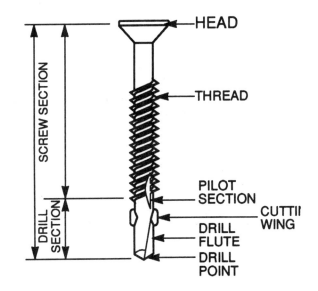

CHECK WITH MANUFACTURER FOR STANDARD AVAILABLE SIZES.
* ⅞" - 1⅝" wood to 18 gauge through 12 gauge.

H.9   FLAT WINGED PHILLIPS   **H.9**

Applications	
Floor Systems	A. Wood Sheathing to Joists
Wall System	A. Wood Sheathing to Studs and Track
Roof Systems	A. Wood Sheathing to Trusses B. Wood Sheathing to Joists / Rafters

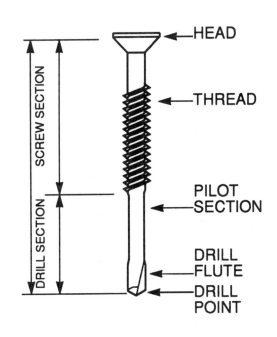

CHECK WITH MANUFACTURER FOR STANDARD AVAILABLE SIZES.
* 1⅛" wood to .210" steel

H.10   PILOT POINT   **H.10**

H. FASTENERS

Applications		
Floor Systems	A. Sheathing to Joist	B. Subflooring to Joist
Wall Systems	A. Sheathing to Stud B. Drywell to Stud	C. Rigid Insulation to Stud
Roof Systems	A. Sheathing to Rafter B. Rigid Insulation to Rafter	C. Ruffing to Rafter D. Drywell to Ceiling Joist

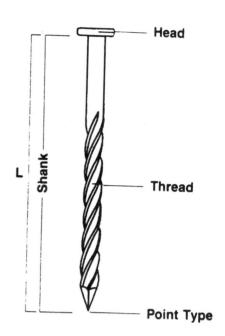

CHECK WITH MANUFACTURER FOR STANDARD AVAILABLE SIZES.

H.11 SCREW-SHANK NAIL

H. FASTENERS

FIBER CEMENT BOARD SCREWS

P/N	BOX QTY.	SIZE		APPLICATION
19300	8M	8x1-1/4	**C-DRILL™**	Designed for fastening fiber cement board to steel studs (20 gauge - 16 gauge). Specially designed wings to allow cutter head to countersink flush with fiber cement board.
19320	5M	8x1-5/8		
19340	2.5M	8x2-1/4		
19400	5M	8x1-1/4	**C-WING™**	Designed for fastening fiber cement board in areas requiring high windshear and pull-through. Recommended for: • Fiber cement board to steel studs • Lap siding • Fiber cement roofing trim • Flooring for tile in kitchen & bath
19420	4M	8x1-5/8		
19440	2.5M	8x2-1/4		
19800	5M	8x1-1/4	**CW-DRILL™**	Designed for fastening fiber cement board to wood. Recommended for: • Fiber cement board to wood • Lap siding • Fiber cement roofing trim • Flooring for tile in kitchen & bath
19820	4M	8x1-5/8		
19840	2.5M	8x2-1/4		
19450	2.5M	12x1-5/8	**CR-DRILL™**	Designed for fastening corrugated fiber cement roofing to steel studs up to 16 gauge or into wood
19460	0.5M	12x3-1/4		
19470	0.5M	12x3-3/4		
19480	0.5M	12x4-3/8	Washers separate in box	
19700	4M	8x1-5/8	**CF-DRILL™**	Special designed Combo Thread fastens flat or corrugated fiber cement fencing to steel posts or girts up to 16 gauge or into wood.
19720	2.5M	8x2-1/4	Washers separate in box	

Illustrations:
- DRILL AND FORM HOLE INTO FIBER CEMENT BOARD WITH THE WINGS
- DRILL AND THREAD INTO STEEL STUD
- COUNTSINK HEAD FLUSH INTO FIBER CEMENT BOARD
- fiber cement roofing

CEMENT BOARD SCREWS

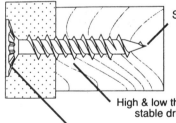

- Sharp point for attaching board to wood stud or metal stud (25-20 gauge)
- High & low thread for stable driving
- Larger wafer head with ribs allows good countersink and clamps the board firmly

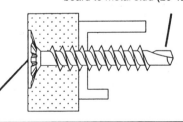

- Self-drilling point for attaching board to metal stud (20-12 gauge)

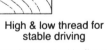

LIFECOAT™

Part No.	Size	No./Ctn.	Wt./Ctn.(lbs)	Applications
CB114D	8X1-1/4	5M	31	For attaching board to metal stud (20-12 gauge)
CB158D	8X1-5/8	4M	31	
CB114S	8X1-1/4	5M	30	For attaching board to wood stud or metal stud (25-20 gauge)
CB158S	8X1-5/8	4M	30	
CB214S	8X2-1/4	2.5M	22	

Information supplied courtesy of Compass International, Buena Park, CA.

H. FASTENERS

SELF-DRILLING SCREWS

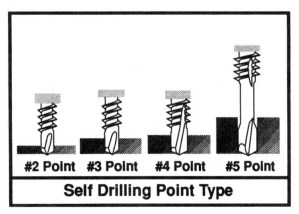

Self Drilling Point Type

Ultimate Self Drilling Screw Strengths
(Figures are approximations for information only. Not accurate for design calculations.)

size diameter	6	8	10	12	14
basic screw diameter (inch)**	.138"	.164"	.190"	.216"	.250"
min. tension (lbs.)	1125 1200	1575 1900	2100 2350 2700	2800 2350	3850 4275
min. torsional strength (in.-lbs.)	25 26	42 50	61 65 75	92 100	150 168
av. shear resistance*	832	1035	1481	2009	2595
av. pullout resistance* (lbs-tension)	619	672	780	800	909
threads per inch**	6-20	8-18	10-24 10-16	12-20 12-14	1/4-20 1/4-14

* 16 Gauge Studs (0.062 inch thick cold rolled steel)
** data apply to most Marker®, RPS®, Deckmark™ and Tapmark® screws.

Material Thickness Recommendations for Standard Self Drilling Screws

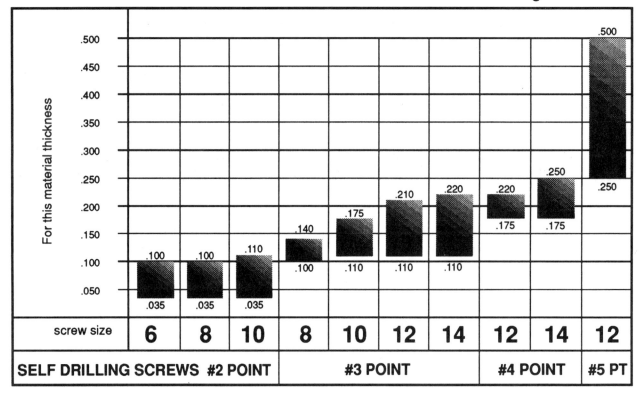

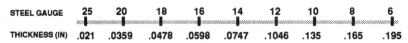

Information supplied courtesy of Compass International, Buena Park, CA.

H. FASTENERS

SPECIFICATION TABLE

Note: The following table of allowable shear and pullout values is based on the procedures prepared by AISI Subcommittee 3 and approved on Feb. 6, 1993 at AISI's Nashville meeting.
Omega = 3.0 S.F.

Gauge of material in contact with screw head
Gauge of material not in contact with screw head

Size No.	Body Dia.	Head O.D.	Spcg.	Gauge 26 / Thk .0179 / Fu 45 shear	26 / .0179 / 45 pull	24 / .0239 / 45 shear	24 / .0239 / 45 pull	22 / .0299 / 45 shear	22 / .0299 / 45 pull	20 / .0359 / 45 shear	20 / .0359 / 45 pull	18 / .0478 / 45 shear	18 / .0478 / 45 pull	16 / .0598 / 65 shear	16 / .0598 / 65 pull	14 / .0747 / 65 shear	14 / .0747 / 65 pull	12 / .1046 / 65 shear	12 / .1046 / 65 pull
#6	.138	.302	.41	56	31	86	42	121	53	159	63	245	84	483	152	603	190	639	266
#8	.164	.322	.49	61	37	94	50	132	63	174	75	267	100	539	181	683	226	683	316
#10	.190	.384	.57	66	43	101	58	142	72	187	87	287	116	580	209	810	261	923	366
#12	.216	.398	.65	70	49	108	66	151	82	199	99	306	132	618	238	863	297	1167	416
1/4"	.250	.480	.75	75	57	116	76	163	95	214	114	329	152	665	275	929	344	1416	482

NOTES:
1. Methods of individual input (calculations) available upon request. Independent lab test results per screw diameter also available upon request.
2. Values derived from independent testing lab results and are specific to compass international products only.
3. If a different manufacturer is used, contact them for their specific calculations and values.

Information supplied courtesy of Compass International, Buena Park, CA.

ALLOWABLE SCREW LOADS

H. FASTENERS

REQUIREMENTS FOR INSTALLING THREADED FASTENERS

ELECTRICAL POWER

ELECTRIC SCREWGUN

- USE HEAVY DUTY SCREWGUN
- 5 AMPS
- 2500 RPM MIN.

DRILL MOTOR

TOO SLOW, BURNS SCREW TIP OFF.

MAGNETIC BIT HOLDER

- #2 Phillips (reduced)
- #3 Phillips
- #1 Square
- #2 Quadrex®

EXTENSION & MAGNETIC HEX SOCKET 1/4", 5/16", 3/8"

BASIC SCREW TYPES

SHARP POINT

- PENETRATES 20 GAUGE OR LESS (THICKNESS).
- REQUIRES PRESSURE TO PENETRATE.
- FASTENS:
  WOOD TO WOOD   STEEL TO STEEL*
  WOOD TO STEEL

  * 2 Layers of 20 GA. steel, use self-drilling screws

SELF-DRILLING

- DRILLS 18 GAUGE OR GREATER (THICKNESS).
- MUST BE ALLOWED TO DRILL WITH EVEN FIRM PRESSURE.
- FASTENS:
  WOOD TO STEEL
  STEEL TO STEEL

GENERAL RULES FOR SELF-DRILLING SCREWS

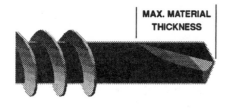

MAX. MATERIAL THICKNESS

- THE MAXIMUM MATERIAL THICKNESS = LENGTH OF FLUTE (DRILL POINT).
- THIS INCLUDES MULTIPLE SHEETS AND GAPS, BUT EXCLUDES PRE-PUNCHED SHEETS
- THREADS CANNOT BE ENGAGED INTO STEEL BEFORE DRILLING OPERATION COMPLETE.

FASTENER HEAD STYLES AND DRIVES

BUGLE PHILLIPS 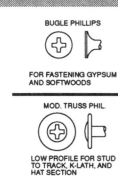 FOR FASTENING GYPSUM AND SOFTWOODS	HEX WASHER FOR STEEL TO STEEL	PAN PHILLIPS FOR STEEL TO STEEL	PAN FRAMING PHILLIPS FOR FRAMING TRACK TO STUD	WAFER PHILLIPS FOR WOOD TO STEEL	QUADREX® NO "CAM-OUT" DRIVE SYSTEM
MOD. TRUSS PHIL. LOW PROFILE FOR STUD TO TRACK, K-LATH, AND HAT SECTION	PANCAKE PHIL. LOW PROFILE FOR STEEL TO STEEL	FLAT PHIL. FOR WOOD TO STEEL	TRIM HEAD SQ. FOR WOOD TO STEEL	HEX WASHER WITH SEALING WASHER FOR EXTERIOR STEEL TO STEEL	

Quadrex is a registered trademark of Isotech Partners, Inc.

Information supplied courtesy of Compass International, Buena Park, CA.

MISC. FASTENER INFORMATION

I. HARDWARE / CONNECTORS

MAS MUDSILL ANCHOR

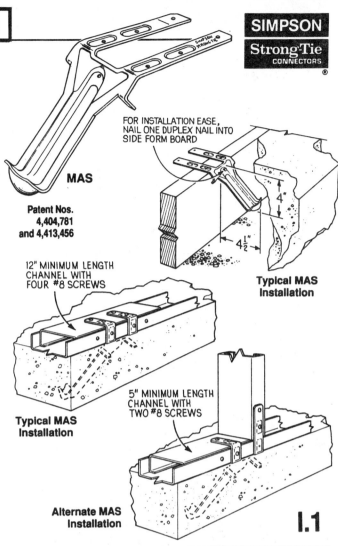

A fast, low installed cost mudsill anchor.

Fast for the finisher—Install before pouring concrete by nailing into form, or insert into concrete after pour. Finish up to edge of slab—no anchor bolts to hand-trowel around, no nuts or washers to lose. For slab or stemwall construction.

MATERIAL: 16 gauge

FINISH: Galvanized. Selected products available with Z-MAX coating; see Corrosion-Resistant Connectors.

INSTALLATION: ▪ Use all specified fasteners. See Screws, page 4.
- Not for use where (a) a horizontal cold joint exists between the slab and foundation wall or footing beneath, unless provisions are made to transfer the load, or (b) anchors are installed in slabs poured over foundation walls formed of concrete block.
- Use a minimum of 2 MAS anchors per track with one MAS located within 1′ from each end of each track.
- Channel section must be attached to the inside of the track for correct MAS installation.

MODEL NO.	FASTENERS		UPLIFT AVG ULT	ALLOWABLE LOADS (133)		
	SIDES TOTAL	TOP		UPLIFT	PARALLEL TO PLATE	PERP TO PLATE
MAS	2-#10	4-#10	2108	845	975	290

1. Loads may not be increased for short-term loading.
2. For alternate installation, uplift load is 585 lbs, parallel-to-plate load is 580 lbs and perpendicular-to-plate is 220 lbs.

I.1

MKP™ MONKEY PAW™ ANCHOR BOLT HOLDER

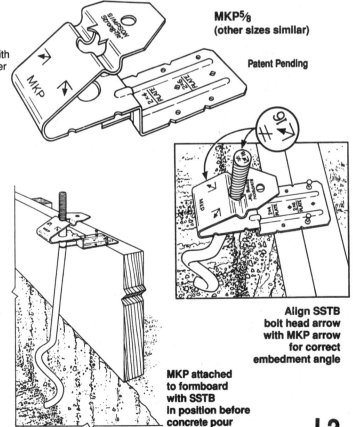

New! Secure the SSTB to the formboard before the concrete pour with a MKP Monkey Paw Bolt Holder. The MKP offers significant savings over plywood or 2x wood holders. Available in ½", ⅝" and ¾" sizes.

SPECIAL FEATURES: ▪ The MKP may be used several times.
- Stabilizes bolt by providing three-point support against lateral concrete pressure.
- Alignment arrows (left or right) match the SSTB bolt head arrow.
- Stamped for 2x4 or 2x6 mudsill placement.
- No nut required to hold bolt in place.
- When removed, the MKP cleans the concrete from the bolt thread.

MATERIAL: 16 gauge

FINISH: Galvanized

INSTALLATION: ▪ Attach SSTB to the MKP. Line up MKP arrows with SSTB bolt head arrows for diagonal installation at approximately 45° from the wall.
- Align bottom tab of MKP with SSTB embedment line.
- Use 2x4 or 2x6 notches to match the mudsill size.
- Attach the MKP to the formboard using duplex nails.
- After concrete pour, allow concrete to cure. Remove the MKP by squeezing the top and bottom together, lifting and twisting off.

I.2

Details and drawings provided courtesy of Simpson Strong-Tie Company, Inc.

I. HARDWARE / CONNECTORS

SIMPSON Strong-Tie® CONNECTORS

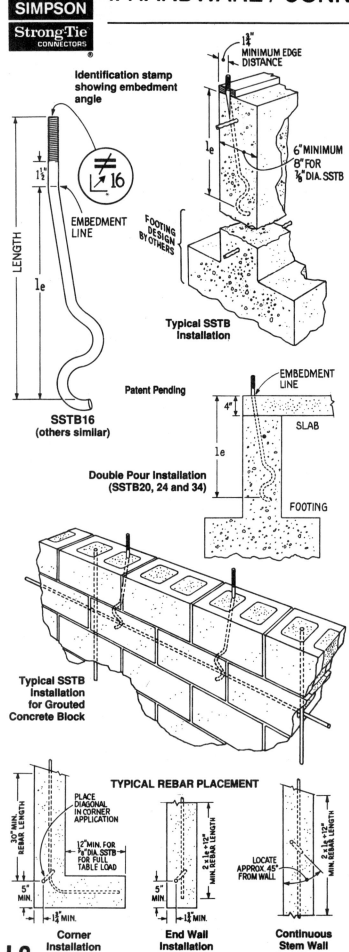

Typical SSTB Installation

Double Pour Installation (SSTB20, 24 and 34)

Typical SSTB Installation for Grouted Concrete Block

TYPICAL REBAR PLACEMENT

Corner Installation — End Wall Installation — Continuous Stem Wall Installation

SSTB® ANCHOR BOLT

The SSTB Anchor Bolt is the first tested and inspection-friendly anchor bolt for Holdowns.

Previously called the STAB, the name was changed to eliminate misinstallation by stabbing the bolt into wet concrete. Extensive testing has been done on the SSTB to determine the design load capacity at a common application, the garage stem wall. The design loads are based on the lowest ultimate, from a series of five tests, with a three times safety factor.

SPECIAL FEATURES:
- Rolled threads for higher tensile capacity.
- Offset angle to reduce side-bursting and provide more concrete cover.
- Stamped bolt head for identification after pour.
- Stamped embedment line to aid installation.
- Configuration results in minimum rebar interference.

INSTALLATION:
- Use the table to select the appropriate SSTB. SSTB is suitable for monolithic and two pour installations.
- Nuts and washers are not supplied with the SSTB; install standard nuts, couplers and/or washers as required.

CONCRETE FOUNDATION
- Install SSTB before the concrete pour using an MKP. Install the SSTB diagonally at approximately 45° from the wall. Install one #4 rebar 3" to 5" from the top of the foundation.
- Minimum concrete compression strength is 2500 psi. Unless noted otherwise, no special inspection is required for foundation concrete when the structural design is based on concrete no greater than 2500 psi (1991 UBC, section 306 (a)1).
- Use 90% of the table load for 2000 psi concrete.

REINFORCED CONCRETE BLOCK
- Install before concrete pour diagonally at approximately 45° in the cell.
- Install one #4 horizontal rebar approx. 12" from the top and #4 vertical rebar minimum 48" o.c.
- Grout all cells with minimum 2000 psi concrete.

OPTIONS: Other SSTB sizes available; contact Simpson for details.

CODE NUMBERS: ICBO No. 4935 and City of L.A. No. RR 25152 for concrete foundation only.

SSTB SELECTION TABLE

MODEL NO.	MONO POUR	TWO POUR
S/HD8	SSTB28	SSTB34
S/HD10	SSTB28	SSTB34

MODEL NO.	DIA	L	MIN EMBED l_e	MAXIMUM ALLOWABLE TENSION LOAD (133)			
				CONCRETE [4,5]		CONCRETE BLOCK [3]	
				EARTHQUAKE	WIND	EARTHQUAKE	WIND
SSTB16	5/8	17	12	4420	3890	4630	4085
SSTB20	5/8	21	16	4600	4050	4630	4085
SSTB24	5/8	25	20	4600	4050	4630	4085
SSTB28	7/8	29	24	10100	8890	—	—
SSTB34	7/8	34	28	10100	8890	—	—
SSTB36	7/8	36	28	10100	8890	—	—

1. Loads may not be increased for short-term loading. Loads apply to wind and earthquake loading per UBC Section 2624 and 2625.
2. Minimum anchor center-to-center spacing is $2l_e$ for anchors acting in tension at the same time for the full load.
3. SSTB with 7/8" dia. have not been tested yet.
4. The maximum allowable load is 8150 lbs. for a SSTB28 used 5" from the end of a concrete foundation. Use the full table load when installed 24" from the end or when installed in the corner condition (see illustration).
5. The SSTB was tested in a stem wall with a minimum amount of concrete cover.

Details and drawings provided courtesy of Simpson Strong-Tie Company, Inc.

HARDWARE / CONNECTORS

ET Epoxy Tie™

Epoxies offer stronger bonding, shorter cure time and less hydrolization than other types of resin anchors. Simpson's ET22 Epoxy-Tie is a two-component amine-based system for high strength anchoring, with a one year shelf life.

Components and Features:
- The ET system has a dual-cartridge, a disposable static mixing nozzle that blends the resin and hardener thoroughly, and a dispensing tool.
- The epoxy is dispensed directly into the anchoring hole, with no waste or mess. ET's unique transparent measuring gauge on the cartridge allows the exact amount to be dispensed.
- The ET is resistant to hydrolization, which occurs when the bond breaks down in the presence of water. The gel consistency allows the material to be injected horizontally as well as vertically.

Installation (see drawings below):
1. Drill hole to specified diameter and depth.
2. Remove dust from hole with manual blower or compressed air. Clean with nylon brush. Dust left in hole will reduce the epoxy's holding capacity.
3. Dispense bead of ET to check for proper mixture, shown by a uniform gray color.
4. Fill hole halfway with ET, starting from the bottom of the hole to avoid air pockets. Withdraw mixing nozzle as the hole is being filled.
5. Insert anchor, turning slowly until the anchor hits the bottom of the hole.

Codes: ICBO # 4945. The ET meets the following specifications: ASTM C 881-90 Standard Specification for Epoxy-Resin-Base Bonding Systems for Concrete. ASTM E 488-90 Standard Test Methods for Strength of Anchors in Concrete and Masonry Elements.

Set Schedule

40°F	60°F	80°F	100°F
18 hrs	6 hrs	4 hrs	4 hrs

Do not disturb anchors during set time.

Cure Schedule

40°F	60°F	80°F	100°F
72 hrs	24 hrs	24 hrs	12 hrs

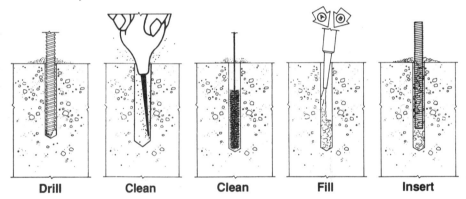

Drill   Clean   Clean   Fill   Insert

ALLOWABLE LOADS FOR THREADED ROD

STUD DIAMETER	DRILL BIT DIAMETER	MINIMUM EMBEDMENT DEPTH	SPACING (s)	EDGE DISTANCE (m)	ALLOWABLE TENSILE LOADS					ALLOWABLE SHEAR LOADS				
					BASED ON BOND STRENGTH		BASED ON STEEL STRENGTH			BASED ON BOND STRENGTH		BASED ON STEEL STRENGTH		
					fc = 2500	fc = 4500	A 307 (SAE 1018)	A 193 Gr. B (SAE 4140)	SS 304	fc = 2500	fc = 4500	A 307 (SAE 1018)	A 193 Gr. B (SAE 4140)	SS 304
3/8"	7/16"	3½"	4½"	2⅝"	2220	2895	2080	4580	1670	1020	1020	1040	2290	1040
½"	9/16"	4¼"	6"	3¼"	2595	4310	3730	8210	2990	2415	2415	1870	4110	1870
5/8"	3/4"	5"	7½"	3¾"	4375	6170	5870	12910	4700	3485	3485	2940	6460	2940
3/4"	7/8"	6¾"	9"	5"	6970	7325	8490	18680	6790	6480	6480	4250	9340	4250
7/8"	1"	7½"	10½"	5⅝"	8005	10640	12000	26400	9500	6240	6720	6000	12800	6000
1"	1⅛"	8¼"	12"	6¼"	10450	12400	15700	34500	12500	7185	7200	7820	17200	7820

1. Allowable loads for bond strength are based on a factor of safety of four on the average ultimate load. They may not be increased for load duration. Allowable load must be the lesser of the bond or steel strength.

2. The tabulated values are for anchors installed at the specified spacing and edge distances. Spacing and edge distances may be reduced in accordance with the table below. Linear interpolation may be used for intermediate spacings.

3. The anchors experience a reduction in tensile and shear capacity with increased ambient temperatures. For reduction values for temperatures above 72°F, consult Simpson's S-ETC brochure.

TENSION CAPACITY		SHEAR CAPACITY		
SPACING (s) AND EDGE DISTANCE (m)	FACTOR (Ft)	EDGE DISTANCE (m)	DIRECTION OF LOAD	FACTOR (Fs)
Spacing min = 0.5s	0.5	0.5m	Toward edge	0.5
Edge distance min = 0.5m	0.5	0.5m	Away from edge	0.5

1. Linear interpolation is allowed for edge distances which fall between 0.5m and 1.0m, and anchor spacing which falls between 0.5s and 1.0s.

Details and drawings provided courtesy of Simpson Strong-Tie Company, Inc.

I. HARDWARE / CONNECTORS

S/PAHD, S/MPAHD, S/HPAHD HOLDOWNS

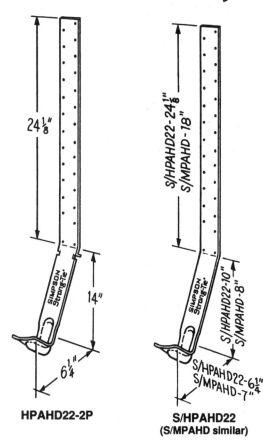

A variety of steel-to-concrete connectors that satisfy engineering and code requirements. Allowable loads include a tested three-times safety factor in concrete. One-piece design; no separate anchors required.

MATERIAL: S/HPA and HPA—10 gauge x 2 1/16"; all others—12 gauge x 2 1/16"

FINISH: Galvanized. Selected products available in Z-MAX coating; see Corrosion-Resistant Connectors.

INSTALLATION: ■ Use all specified fasteners. See Screws, page 4.
  ■ Unless otherwise noted, do NOT install where:
    (a) a horizontal cold joint exists within the embedment depth between the slab and foundation wall or footing beneath, unless provisions are made to transfer the load, or the slab is designed to resist the load imposed by the anchor; or
    (b) slabs are poured over concrete block foundation walls.
  ■ To get the full table load, the minimum center-to-center spacing is twice the embedment depth when resisting tension loads at the same time.
  ■ **FOUNDATION CORNERS:** Screw quantities have been reduced when the load is limited by tested concrete pullout strength. Additional screw holes need not be filled.
  ■ Loads are calculated using a 4" slab over 8" and 10" foundation walls.

S/PAHD42, S/MPAHD, S/HPAHD22, HPAHD22-2P HOLDOWNS:
  Designed to be installed at the edge of concrete. Tests determined the pullout strength with one horizontal #4 rebar in the shear cone.
  Install before pouring concrete by nailing to the form. Pre-bent to control the embedment at the required angle; field-bending is not necessary.
  Installation holes allow nailing to the form, resulting in 1" deeper embedment; see illustration.

OPTIONS: See also S/HD Holdowns, S/LTT and S/MTT Tension Ties.

HPAHD22-2P

S/HPAHD22
(S/MPAHD similar)

Typical HPAHD22-2P before Concrete Pour (S/MPAHD and S/HPAHD similar)

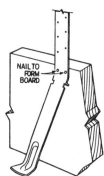

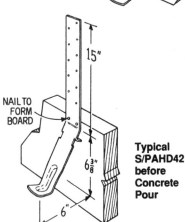

Typical S/PAHD42 before Concrete Pour

MODEL NO.	MINIMUM FOOTING WIDTH	SCREWS	ALLOWABLE LOADS (133)
EDGE INSTALLATION—2500 PSI CONCRETE			
SINGLE POUR—see installation 1—8" min from corner			
S/PAHD42	6	7- #10	2205
	8	9- #10	2945
S/MPAHD	6	9- #10	2800
	8	12- #10	3665
S/HPAHD22	6	10- #10	3150
	8	14- #10	4370
DOUBLE POUR—see installation 3—8" min from corner			
S/PAHD42	6	7- #10	2205
	8	7- #10	2255
S/MPAHD	6	9- #10	2800
	8	12- #10	3665
S/HPAHD22	6	10- #10	3150
	8	12- #10	3950
HPAHD22-2P	6	10- #10	3150
	8	14- #10	4415

MODEL NO.	MINIMUM FOOTING WIDTH	SCREWS	ALLOWABLE LOADS (133)
CORNER INSTALLATION—2000 PSI CONCRETE			
SINGLE POUR—see installation 2—1/2" min from corner			
S/PAHD42	6	4- #10	1225
	8	5- #10	1400
S/MPAHD	6	4- #10	1135
	8	6- #10	1840
S/HPAHD22	6	6- #10	1750
	8	7- #10	2210
DOUBLE POUR—see installation 4—1/2" min from corner			
S/PAHD42	6	4- #10	1225
	8	5- #10	1400
S/MPAHD	6	4- #10	1135
	8	6- #10	1840
S/HPAHD22	6	6- #10	1750
	8	7- #10	2210
HPAHD22-2P	6	7- #10	2210
	8	7- #10	2210

1. S/HPAHD22 may be embedded 4" into the slab and 6" into the 8" footing beneath for a maximum load of 2810 lbs. at 8" minimum from the closest corner, and 1400 lbs. at 1/2" from the closest corner.

2. EDGE INSTALLATION: The minimum concrete compression strength is 2500 psi. For 2000 psi, calculate loads at 0.75 of the table allowable loads. CORNER INSTALLATION: The minimum concrete compression strength is 2000 psi. No load reduction is allowed.

3. Allowable loads have been increased 33% for wind or earthquake loading with no further increase allowed.

4. Calculate the loads using straight line interpolation for corner distances between 1/2" and 8".

Details and drawings provided courtesy of Simpson Strong-Tie Company, Inc.

HARDWARE / CONNECTORS

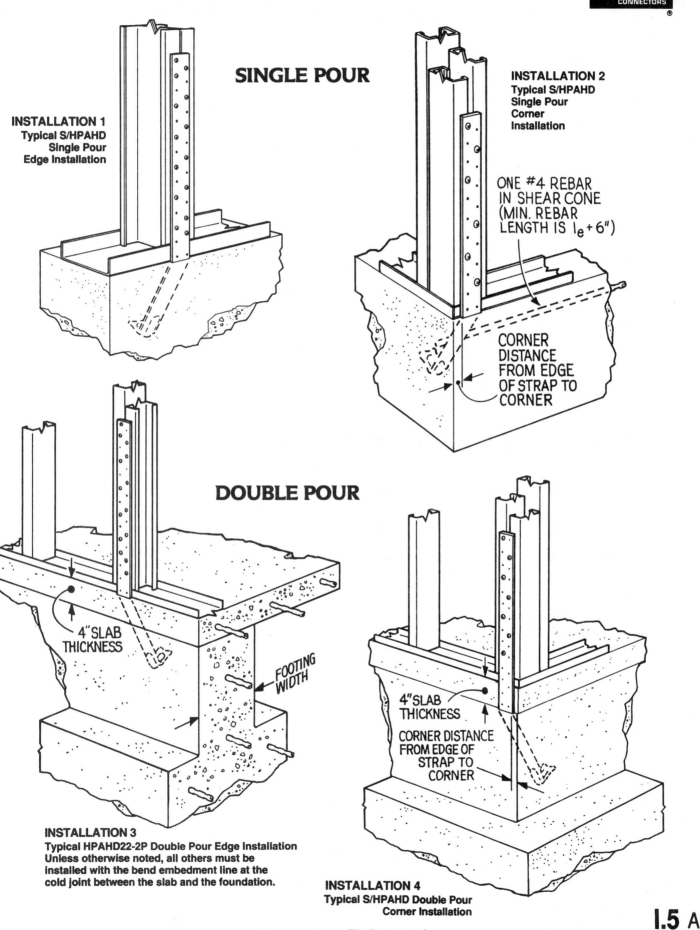

I.5 A

I. HARDWARE / CONNECTORS

S/HD HOLDOWNS

The S/HD's design makes installation easy. The connector height does not interfere with stud knockouts, and the narrow width fits within the stud section.

MATERIAL: S/HD8—10 ga with ¼" plate; S/HD10—10 ga with ⅜" plate
FINISH: Simpson gray paint
INSTALLATION: ■ Use all specified fasteners. See Screws, page 4.
■ See SSTB Anchor Bolts. The design engineer may specify any alternate anchorage calculated to resist the tension load for your specific job.

MODEL NO.	DIMENSIONS			FASTENERS		AVG ULT	ALLOWABLE LOAD
	W	H	CL	ANCHOR DIA	SCREWS		
S/HD8	2½	13⅞	1½	⅞	24- #10	21167	7920
S/HD10	2½	16⅛	1½	⅞	30- #10	29000	9900

1. Specify the anchor embedment and configuration. See SSTB Anchor Bolts.
2. Allowable loads have been increased 33% for wind or earthquake loading with no further increase allowed; reduce where other load durations govern.

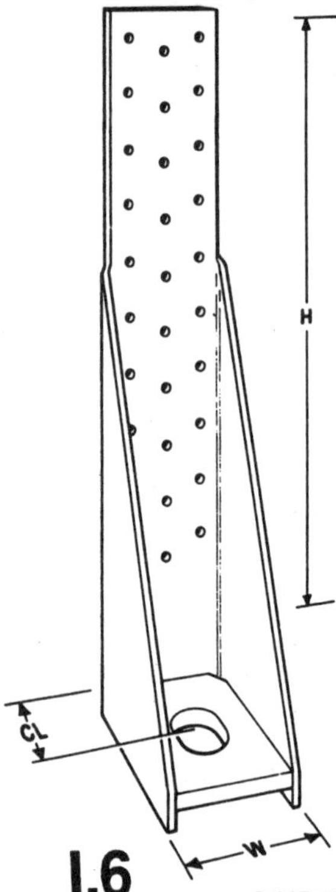

Washers are not required for S/HD10.

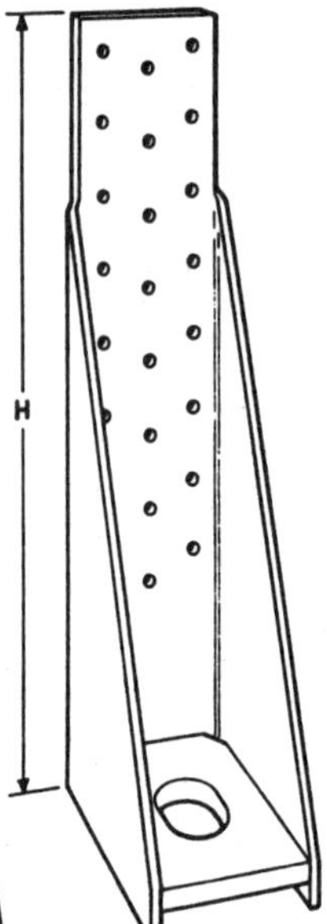

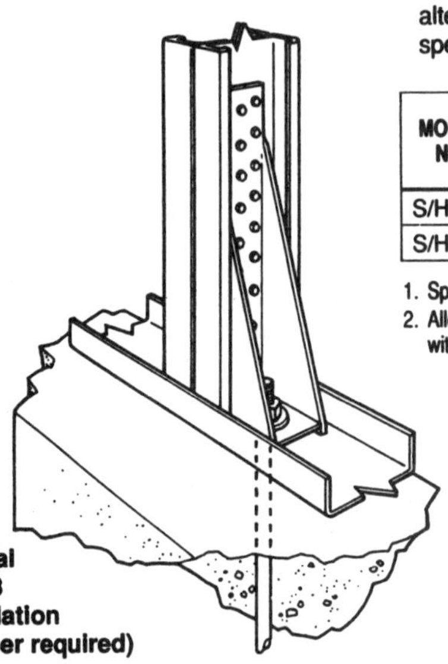

Typical S/HD8 Installation (washer required)

S/HD8
S/HD10

S/LTT, S/MTT, S/HTT TENSION TIES

The S/MTT14 and S/HTT14 are single-piece formed tension ties—no rivets, and a 4-ply formed seat which won't unfold during loading. No washers are required.

The S/LTT and S/MTT Tension Ties are ideal for retrofit or new construction projects. They provide high strength, post-pour, concrete-to-steel connections.

MATERIAL: See table
FINISH: Galvanized.
INSTALLATION: ■ Use all specified fasteners. See Screws, page 4.
■ Use the specified number and type of screws to attach the strap portion to the steel stud. Bolt the base to the wall or foundation with a suitable anchor; see table for the required bolt diameter.
■ The S/MTT14 and S/HTT14 can have a maximum offset of 2" from the stud face to the centerline of the anchor bolt.

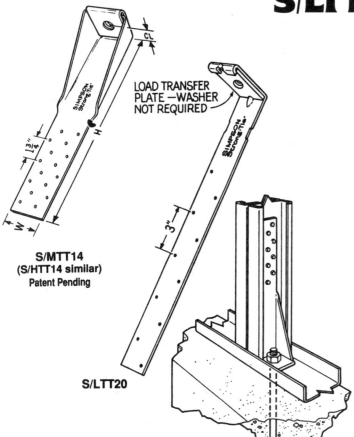

S/MTT14 (S/HTT14 similar) Patent Pending

LOAD TRANSFER PLATE—WASHER NOT REQUIRED

S/LTT20

Typical S/HTT14 Installation as a Holdown

MODEL NO.	MATERIAL		DIMENSIONS			FASTENERS		ALLOWABLE LOADS (133)
	STRAP	PLATE	W	H	CL	ANCHOR BOLTS	SCREWS	
S/LTT20	12 ga	3 ga	2	20	1½	½	6- #10	1750
S/MTT14	12 ga	—	2½	15	1 1/16	⅝	14- #10	4620
S/HTT14	11 ga	—	2½	15	1 1/16	⅝	16- #10	5260

1. The designer may specify anchor bolt type, length and embedment.
2. Allowable loads have been increased 33% for wind or earthquake loading with no further increase allowed.

Details and drawings provided courtesy of Simpson Strong-Tie Company, Inc.

HARDWARE / CONNECTORS

W, WNP HANGERS

This series has the greatest design flexibility and versatility. The hanger's straight side-flanges support the top and bottom of the channel for a strong, balanced connection.

MATERIAL: Stirrup—12 gauge

FINISH: Simpson gray paint. Some models available hot-dipped galvanized; specify HDG.

INSTALLATION: Hangers may be welded to steel headers with ⅛" for W and 3/16" for WNP by 1½" fillet welds located at each end of the top flange.

OPTIONS: ■ W and H dimensions are modifiable.

SLOPED AND/OR SKEWED SEAT

- W/WNP series may be skewed to a maximum of 84° and/or sloped to a maximum of 45°.
- For slope only, skew only, or slope and skew combinations, the allowable load is 100% of the table load.
- Specify the slope up or down in degrees from the horizontal plane and/or the skew right or left in degrees from the perpendicular vertical plane. Specify whether low side, high side or center of joist will be flush with the top of the header.

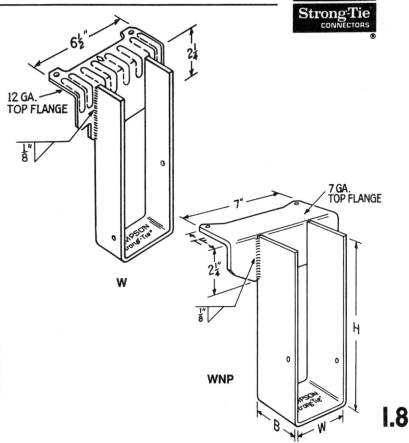

MODEL NO.	DIMENSIONS			FASTENERS		ALLOWABLE LOADS
	W	H	B	HEADER	JOIST	
W	1 9/16 - 7½	3½ - 30	2 - 6	Weld	2-#10	2200
WNP	1 9/16 - 7½	3½ - 30	2 - 6	Weld	2-#10	3255

I.8

LB, B HANGERS

Precision forming with manufacturing quality control provides dimensional accuracy and helps ensure proper bearing area and connection. These designs have the material section where it counts, resulting in maximum loads.

MATERIAL: LB—14 gauge; B—12 gauge

FINISH: Galvanized

INSTALLATION: ■ **LB** may be used for weld-on applications; a minimum of 2" x material thickness of weld on each top flange is required. Distribute the weld equally on both top flanges. Consult the code for special considerations when welding galvanized steel. Uplift loads do not apply to weld-on applications.

■ **B** may be used for weld-on applications. The minimum required weld to the top flanges is ⅛" x 2" fillet weld to each side of each top flange tab. Distribute the weld equally on both top flanges. Uplift loads do not apply to weld-on applications.

OPTIONS: B series can be sloped to a maximum of 45°. For 0° to 30°, the allowable load is 100% of the table load. For 31° to 45°, the maximum allowable download is 80% of the table roof load.

CODE NUMBER: ICBO No. 1258.

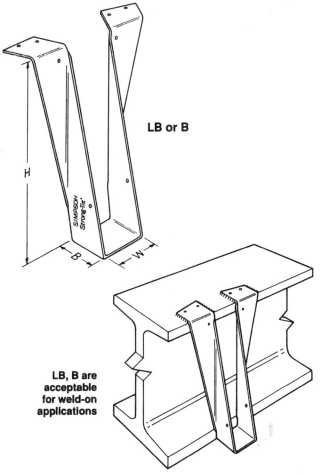

LB or B

LB, B are acceptable for weld-on applications

MODEL NO.	DIMENSIONS			FASTENERS		ALLOWABLE LOADS
	W	H	B	HEADER	JOIST	
LB	1 9/16 - 3 9/16	3½ - 20	2 - 3	Weld	2-#10	1550
B	1 9/16 - 7½	7 - 30	2 - 3	Weld	2-#10	2415

I.9

Details and drawings provided courtesy of Simpson Strong-Tie Company, Inc.

I. HARDWARE / CONNECTORS

L, S/LS REINFORCING AND SKEWABLE ANGLES

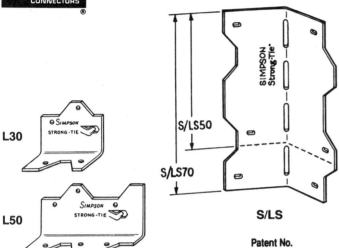
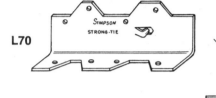
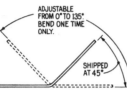

General utility reinforcing angles with multiple uses.
S/LS— Field-adjustable angles attach members intersecting at angles.

MATERIAL: L—16 gauge; S/LS—18 gauge
FINISH: Galvanized
INSTALLATION: ■ Use all specified fasteners. See Screws, page 4.
- S/LS—field-skewable; bend one time only.
- Joist must be constrained against rotation when using a single S/LS per connection.

MODEL NO.	LENGTH	FASTENERS	ALLOWABLE LOADS F_1	F_2
L30	3	4- #10	255	60
L50	5	6- #10	965	110
L70	7	8- #10	1375	100
S/LS50	$4^7/_8$	4- #10	600	—
S/LS70	$6^3/_8$	6- #10	915	—

1. No load duration increase allowed.
2. Loads are for one part only.
3. L30 loads are based on 20 gauge and heavier members. All other loads are based on 16 gauge and heavier members.

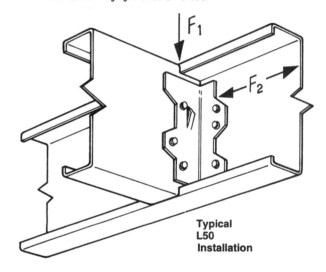

Typical L30 Installation

Typical L50 Installation

I.10

A, S/A ANGLES

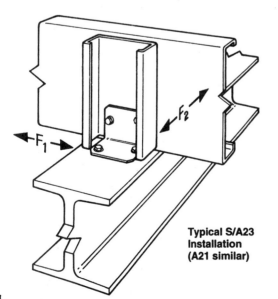

Typical S/A23 Installation (A21 similar)

MATERIAL: 18 gauge
FINISH: Galvanized
INSTALLATION: Use all specified fasteners. See Screws, page 4.

MODEL NO.	DIMENSIONS W_1	W_2	L	FASTENERS	ALLOWABLE[1] LOADS F_1	F_2
A21	2	$1^1/_2$	$1^3/_8$	4- #10	150	50
S/A23	2	$1^1/_2$	$2^3/_4$	4- #10	310	70

1. No load duration increase allowed.

Details and drawings provided courtesy of Simpson Strong-Tie Company, Inc.

SIMPSON Strong-Tie CONNECTORS ®

S/H SEISMIC AND HURRICANE TIES

Designed to provide wind and seismic ties for trusses
d rafters, this versatile line may be used for general tie
rposes, strongback attachments, and as all-purpose
s where one member crosses another.

ATERIAL: 18 gauge

NISH: Galvanized. Selected products available in stainless steel or Z-MAX coating; see Corrosion-Resistant Connectors.

STALLATION:
- Use all specified fasteners. See Screws, page 4.
- The S/H1 can be installed with flanges facing outwards (reverse of illustration #1). When installed inside a wall for truss applications.
- Ties are shipped in equal quantities of separate rights and lefts.
- S/H1 does not replace solid blocking.

ODEL NO.	FASTENERS			MAX ALLOWABLE LOADS		
	TO RAFTERS	TO PLATES	TO STUDS	UPLIFT (133)	LATERAL	
					F_1 (133)	F_2 (133)
/H1	3- #10	2- #10	1- #10	330	100	115
/H2	3- #10	—	3- #10	395	—	—
/H2.5	4- #10	—	4- #10	415	90	125
/H3	2- #10	2- #10	—	380	90	125

Loads have been increased 33% for wind or earthquake loading; no further increase allowed.

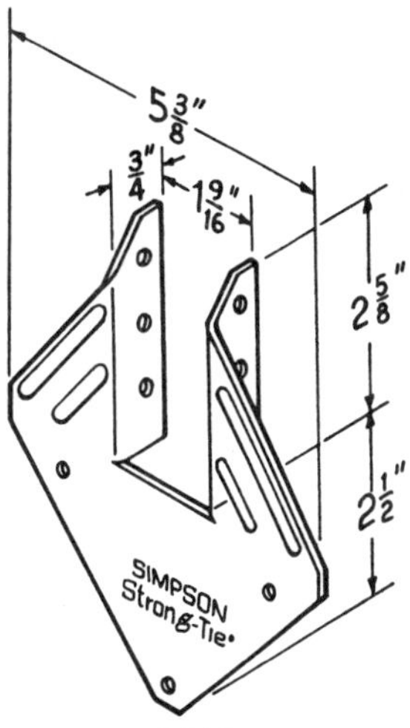

S/H1

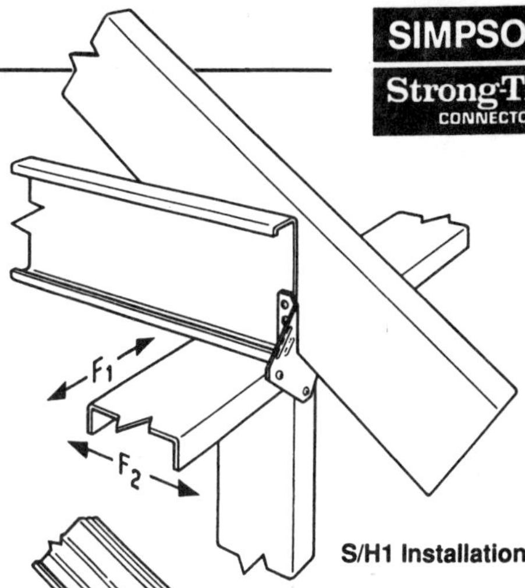

S/H1 Installation

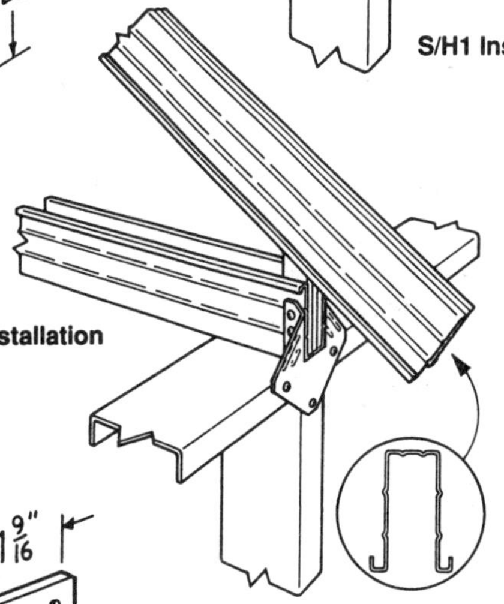

S/H1 Installation

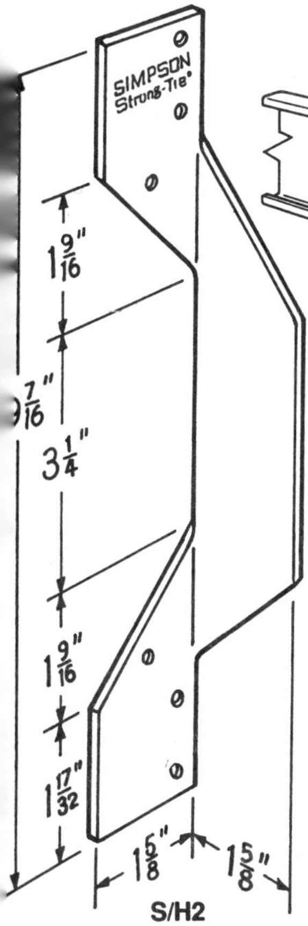

S/H2

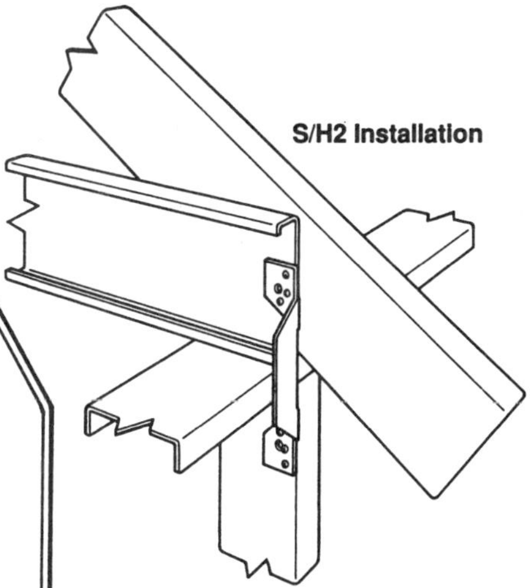

S/H2 Installation

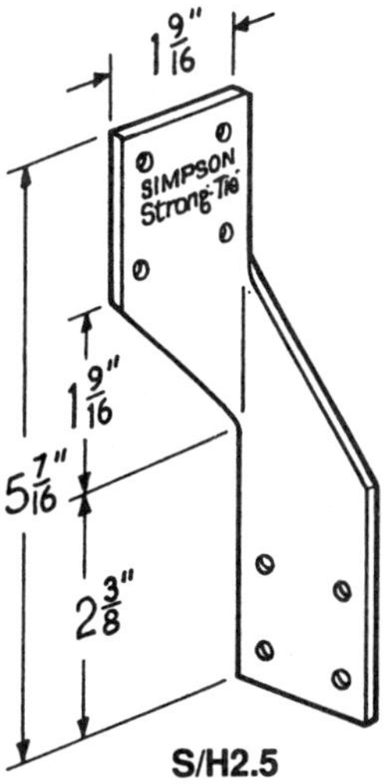

S/H2.5

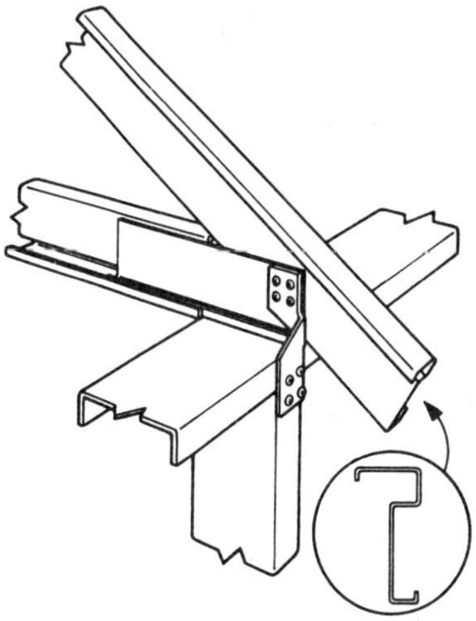

S/H2.5 Installation

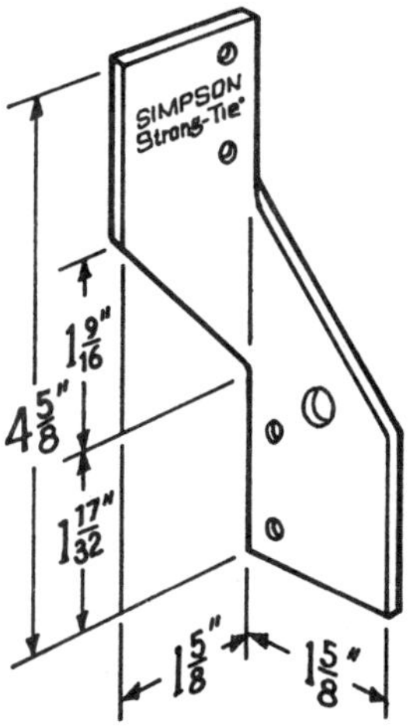

S/H3

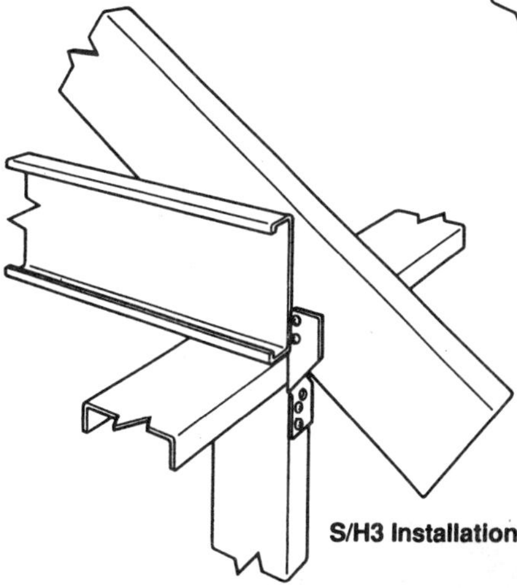

S/H3 Installation

s and drawings provided courtesy of Simpson Strong-Tie Company, Inc.

I. HARDWARE / CONNECTORS

LTS, MTS TWIST STRAPS

Twist straps provide a tension connection between two members. These 1¼" wide straps are an economical way to resist uplift at the heel of a truss.

The 3" bend section eliminates interference at the transition points.

MATERIAL: MTS—16 gauge; LTS—18 gauge
FINISH: Galvanized. Selected products available in stainless steel and Z-MAX coating; see Corrosion-Resistant Connectors.
INSTALLATION: Use all specified fasteners. See Screws, page 4.

MODEL NO.	LENGTH	FASTENERS (TOTAL)	ALLOWABLE LOADS (133)
LTS8	8	6- #10	400
MTS8	8	6- #10	640

1. Install half of the fasteners on each end of the strap to achieve full loads.
2. Loads have been increased 33% for wind or earthquake loading with no further increase allowed.

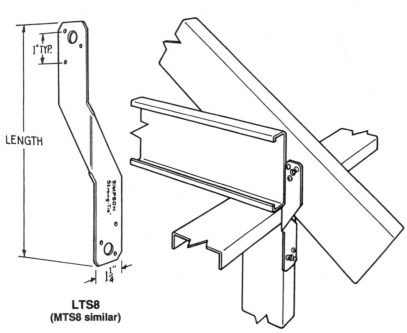

LTS8 (MTS8 similar)

Typical LTS Installation Truss to Steel Studs

I.13

ST, S/MST STRAP TIES

Install Strap Ties where plates or soles are cut, at wall intersections, floor-to-floor applications, and as ridge ties and truss plates.
FINISH: Galvanized.
INSTALLATION: Use all specified fasteners. See Screws, page 4.

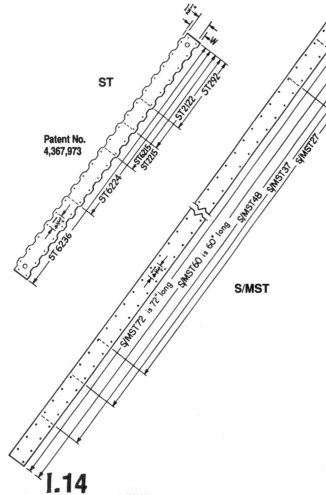

Patent No. 4,367,973

MODEL NO.	MATL	DIMENSIONS W	DIMENSIONS L	FASTENERS (TOTAL)	ALLOWABLE LOADS (133)
ST292	20 ga	2 1/16	9 5/16	8- #10	1075
ST2122	20 ga	2 1/16	12 13/16	10- #10	1425
ST2115	20 ga	3/4	16 5/16	4- #10	600
ST2215	20 ga	2 1/16	16 5/16	10- #10	1615
ST6215	16 ga	2 1/16	16 5/16	12- #10	1785
ST6224	16 ga	2 1/16	23 5/16	16- #10	2500
ST6236	14 ga	2 1/16	33 13/16	22- #10	3300
S/MST27	12 ga	2 1/16	27	18- #10	2675
S/MST37	12 ga	2 1/16	37	24- #10	3745
S/MST48	12 ga	2 1/16	48	28- #10	4460
S/MST60	10 ga	2 1/16	60	36- #10	5800
S/MST72	10 ga	2 1/16	72	36- #10	5800

1. Maximum loads have been increased for wind or earthquake loading with no further increase allowed.

I.14

Details and drawings provided courtesy of Simpson Strong-Tie Company, Inc.

I. HARDWARE / CONNECTORS

CS, CMST COILED STRAPS

CS are continuous utility straps which can be cut to length on the job site. Packaged in a lightweight (about 40 pounds), portable 2' square carton. The popular 18 gauge strap is available in 100' or 200' rolls—specify CS18-100 or CS18-200.

MATERIAL: See table
FINISH: Galvanized. Selected products available in Z-MAX; see Corrosion-Resistant Connectors.
INSTALLATION: ▪ Use all specified fasteners. See Screws pg 4.
▪ The table shows the maximum allowable loads and the screws required to obtain them. Fewer screws may be used; reduce the allowable load by the code lateral load for each screw subtracted from each end.

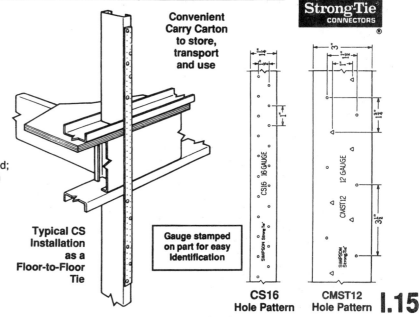

Typical CS Installation as a Floor-to-Floor Tie

Convenient Carry Carton to store, transport and use

Gauge stamped on part for easy Identification

CS16 Hole Pattern

CMST12 Hole Pattern

MODEL NO.	MATERIAL	TOTAL LENGTH	TOTAL FASTENERS	ALLOWABLE LOADS (133)
CMST12	12 ga	40'	60- #10	9640
CS16	16 ga	150'	12- #10	1650
CS18	18 ga	100' & 200'	8- #10	1270
CS20	20 ga	250'	8- #10	1005
CS22	22 ga	300'	6- #10	825

1. 133% value may be used for wind or earthquake loading.

PSC PLYWOOD SHEATHING CLIPS

MATERIAL: 18 gauge
FINISH: Galvanized
INSTALLATION: Models (sizes) available are 3/8 PSC, 7/16 PSC, 15/32 PSC, 1/2 PSC, 19/32 PSC, 5/8 PSC, and 3/4 PSC.

SPAN RATING	PLYWOOD THICKNESS	MAXIMUM SPAN		PSCs PER SPAN
		WITH PSCs	WITHOUT PSCs	
24/0	3/8, 7/16	24	20	1
32/16	15/32, 1/2, 5/8	32	28	1
40/20	19/32, 5/8, 3/4	40	32	1
48/24	3/4	48	36	2

1. Span ratings for APA Rated Sheathing when the long dimension or strength axis is across three or more supports.

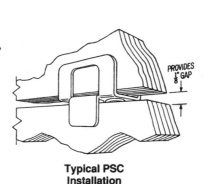
Typical PSC Installation

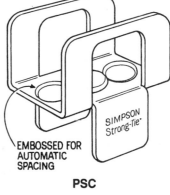
PSC

TB, LTB BRIDGING

TB—Tension-type bridging with maximum fastener flexibility. Use two of the seven screw holes at each end.

LTB—Staggered fastener pattern accommodates 6" to 12" web height. Use two of the holes at each end.

MATERIAL: LTB—22 gauge; TB—20 gauge
FINISH: Galvanized
INSTALLATION: Bridging will fit flange widths from 1 5/8 to 3".

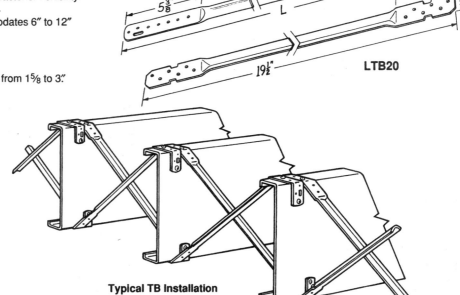

WEB HEIGHT	SPACING	TB		LTB
		MODEL NO.	L	MODEL NO.
6"	12" o.c.	TB20	20	LTB20
8"	12" o.c.	TB20	20	LTB20
10"	12" o.c.	TB20	20	LTB20
12"	12" o.c.	TB27	27	—
6"	16" o.c.	TB27	27	—
8"	16" o.c.	TB27	27	—
10"	16" o.c.	TB27	27	—
12"	16" o.c.	TB27	27	—
6"	24" o.c.	TB36	36	—
8"	24" o.c.	TB36	36	—
10"	24" o.c.	TB36	36	—
12"	24" o.c.	TB36	36	—

Typical TB Installation

Details and drawings provided courtesy of Simpson Strong-Tie Company, Inc.

I. HARDWARE / CONNECTORS

U.S. Standard Steel Gauge Equivalents in Nominal Dimensions

GAUGE	APPROXIMATE DIMENSIONS		DECIMALS (INCHES)		
	in	mm	UNCOATED STEEL	GALVANIZED STEEL (G60)	Z-MAX™
3	¼	6	0.239"	—	—
7	³⁄₁₆	4.5	0.179"	—	—
10	⁹⁄₆₄	3.4	0.134"	0.138"	0.140"
11	⅛	3	0.120"	0.123"	0.125"
12	⁷⁄₆₄	2.7	0.105"	0.108"	0.110"
14	⁵⁄₆₄	2	0.075"	0.078"	0.080"
16	¹⁄₁₆	1.5	0.060"	0.063"	0.065"
18	³⁄₆₄	1.2	0.048"	0.052"	0.054"
20	¹⁄₃₂	1	0.036"	0.040"	0.042"
22	¹⁄₃₂	0.8	0.030"	0.034"	0.036"

1. Actual steel dimensions will vary from nominal dimensions according to industry tolerances.

Bolt Diameter Conversion

in	mm
⅜	9.5
½	12.7
⅝	15.9
¾	19.1
⅞	22.2
1	25.4

Metric Conversion Chart

IMPERIAL	METRIC
1 in	25.40 mm
1 ft	0.3048 m
1 lb	4.448 N
1 Kip	4.448 kN
1 psi	6895 Pa

mm = millimeter
m = meter
N = newton
kN = kilonewton
Pa = pascal

Roof Slope Conversion Chart

RISE/RUN	SLOPE
1/12	5°
2/12	10°
3/12	14°
4/12	18°
5/12	23°
6/12	27°
7/12	30°
8/12	34°
9/12	37°
10/12	40°
11/12	42°
12/12	45°

Slope rounded to the nearest degree.

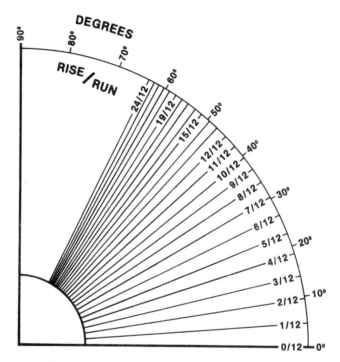

NOW AVAILABLE!

STEEL FRAMING BLUEPRINTS

Complete sample reference blueprints for single story and two story steel framed designed homes. An Excellent source for architects, engineers and contractors looking for an easy reference to the construction of steel framing in single family construction. These plans utilize the details contained in the **Residential Steel Framing Construction Guide** and provide the guidelines and format necessary for building department approvals.

Each plan comes with wall panel layout, second floor framing plan (two story), roof framing plan and corresponding detail sheets. Plans are available in half size sheets for quick reference and easy storage in your briefcase or desk.

Single story - $19.95 plus $5.00 shipping and handling (add $1.50 shipping for additional plan shipped) Mainland U.S.

Two story - $24.95 plus $5.00 shipping and handling (add $1.50 shipping for each additional plan shipped) Mainland U.S.

Note: Each plan has been through plan check and received Building Dept. approval. A 10% discount is available when ordering both plans.
* Full size sheets are available for an additional $8.00 per plan.
** Additional shipping charges for overseas or out of country delivery.

ORDER FORM

Name: _____ Address: _____ City: _____
State: _____ Zip Code: _____ Tel. # ( ) _____ Credit Card # _____
Mcard _____ Visa _____ Exp. date _____ Quantity Ordered: _____
Single story _____ Two story _____

Send check or money order to Technical Publications 3020 Builders Ave. Las Vegas, NV. 89101. California residents add 7.75% sales tax. Prices and availability are subject to change without notice. Visa and Mastercard accepted. Allow up to 4 weeks delivery.

Also Available Through Technical Publications:

— Consulting
— Engineering / Design
— Steel Framing Products / Manufacturers
— Truss Design
— Drafting
— Architectural Services
— Spanish version of *Residential Steel Framing Construction Guide*

It is the intent of the author to provide the above services at his discretion whenever possible for a fee.

The author reserves the right to provide or not provide these services at any time.

If you are interested in any of these additional products or services please check the appropriate box and send this form to:

Technical Publications
c/o Steel Services
3020 Builders Ave.
Las Vegas, NV 89101

Company Name or Person: _____
Contact: _____
Address: _____
Phone: _____
Fax: _____
Project type (circle one): Custom Home, Multi-family, Light commercial, Production (tract) Homes, Other: _____
Project location (city, state, country): _____
Start date: _____
Additional Information: _____
_____
_____